T0135167

Studies in Big Data

Volume 29

About this Series

The series "Studies in Big Data" (SBD) publishes new developments and advances in the various areas of Big Data-quickly and with a high quality. The intent is to cover the theory, research, development, and applications of Big Data, as embedded in the fields of engineering, computer science, physics, economics and life sciences. The books of the series refer to the analysis and understanding of large, complex, and/or distributed data sets generated from recent digital sources coming from sensors or other physical instruments as well as simulations, crowd sourcing, social networks or other internet transactions, such as emails or video click streams and other. The series contains monographs, lecture notes and edited volumes in Big Data spanning the areas of computational intelligence incl. neural networks, evolutionary computation, soft computing, fuzzy systems, as well as artificial intelligence, data mining, modern statistics and Operations research, as well as self-organizing systems. Of particular value to both the contributors and the readership are the short publication timeframe and the world-wide distribution, which enable both wide and rapid dissemination of research output.

More information about this series at http://www.springer.com/series/11970

L. Octavio Lerma · Vladik Kreinovich

Towards Analytical Techniques for Optimizing Knowledge Acquisition, Processing, Propagation, and Use in Cyberinfrastructure and Big Data

L. Octavio Lerma
Department of Computer Science, College of Engineering
The University of Texas at El Paso
El Paso, TX
USA

Vladik Kreinovich
Department of Computer Science, College of Engineering
The University of Texas at El Paso
El Paso, TX
USA

ISSN 2197-6503 ISSN 2197-6511 (electronic)
Studies in Big Data
ISBN 978-3-319-87058-8 ISBN 978-3-319-61349-9 (eBook)
DOI 10.1007/978-3-319-61349-9

Softcover reprint of the hardcover 1st edition 2017

Printed on acid-free paper

This Springer imprint is published by Springer Nature
The registered company is Springer International Publishing AG
The registered company address is: Gewerbestrasse 11, 6330 Cham, Switzerland

Preface

Motivation. For many decades, there has been a continuous progress in science and engineering applications. A large part of this progress comes from the new knowledge that researchers acquire, propagate, and use. This new knowledge has revolutionized many aspects of our life, from driving to communications to shopping.

Somewhat surprisingly, there is one area of human activity which is the least impacted by the modern technological progress: the very processes of acquiring, processing, and propagating information. When we decide where to place sensors, which algorithm to use for processing the data—we rely mostly on our own intuition and on the opinion of the experts. As a result, knowledge-related methods that we select are often far from optimal. To make effective recommendations, it is necessary to build realistic models of the corresponding processes and then use these models to find optimal ways of controlling these processes.

The need for such models is well understood. There are many numerical models of knowledge acquisition, processing, and propagations. Some of these models have been successfully used to enhance the corresponding processes. However, these applications are limited by the fact that most of these models are based on detailed numerical simulation of the corresponding processes, which make the resulting models very time-consuming to use. This is especially important in situations of *big data*, when the amount of data is so huge that the traditional numerical methods are not applicable. It is therefore necessary to develop *analytical* models for the corresponding knowledge-related processes, models that would allow easier optimization and application.

Structure of the Book. The main purpose of this book is to develop analytical models for all the knowledge-related processes, from knowledge acquisition to knowledge processing and knowledge propagation. In this book:

- in Chap. 2, we provide analytical models for data acquisition, i.e., for the use of sensors;

- in Chap. 3, we provide analytical models for data and knowledge processing; specifically, we explain how to best organize computing power, how to best organize research, and how to best organize research teams;
- in Chap. 4, we provide analytical models for knowledge propagation and resulting knowledge enhancement; these models describe how the knowledge propagates, how to assess the students' initial knowledge level, how to present the material, and what is the effect of feedback;
- finally, in Chap. 5, we provide case studies explaining how the corresponding models can be used in science, in control, in design, and in maintenance.

Acknowledgements. Our thanks to Dr. Miguel Argaez, Dr. Ann Gates, Dr. Olga Kosheleva, Dr. Ming-Ying Leung, Dr. Deana Pennington, Dr. Andrzej Pownuk, Dr. Scott Starks, Dr. Craig Tweedie, and Dr. Leticia Velasquez for their help, valuable suggestions, and inspiring ideas. We would also like to express our appreciation to Paul Salopek and his Out of Eden Walk Project team for their help.

Thanks as well to all the participants of:

- 8th International Joint Conference on Autonomous Agents and Multiagent Systems AAMAS'09 (Budapest, Hungary, May 10–15, 2009),
- 30th Annual Conference of the North American Fuzzy Information Processing Society NAFIPS'2011 (El Paso, Texas, March 18–20, 2011),
- World Conference on Soft Computing (San Francisco, CA, May 23–26, 2011),
- AAAI Workshop on Applied Adversarial Reasoning and Risk Modeling AARM'11 (San Francisco, California, August 7, 2011),
- 11th International Sun Conference on Teaching and Learning (El Paso, Texas, March 5–7, 2014),
- International Conference on Risk Analysis in Meteorological Disasters RAMD'2014 (Nanjing, China, October 12–13, 2014),
- Annual Conference of the North American Fuzzy Information Processing Society NAFIPS'2015 and 5th World Conference on Soft Computing (Redmond, Washington, August 17–19, 2015),

for valuable discussions.

Grant support. This work was supported in part by the US National Science Foundation Grants:

- Grants No. HRD-0734825 and HRD-1242122 (Cyber-ShARE Center of Excellence),
- Grant No. DUE-0926721 (Bioinformatics), and
- Grant No. OCI-1135525 (CI-Team Diffusion project: The Virtual Learning Commons).

El Paso, TX, USA
August 2017

L. Octavio Lerma
Vladik Kreinovich

Contents

Chapter 1
Introduction

Abstract In this chapter, we explain what is the main problems with which we deal in this book, how exactly we deal with them, and what is the book's structure.

Knowledge-related processes are important. For many decades, there has been a continuous progress in science and engineering applications. A large part of this progress comes from the new knowledge that researchers acquire, propagate, and use. This new knowledge has revolutionized many aspects of our life. We rely on automated computer-based systems when we drive our cars, when we fly planes, when we communicate with people over the web.

Knowledge-related processes need to be enhanced. Somewhat surprisingly, there is one area of human activity which is the least impacted by the modern technological progress: the very processes of acquiring, processing, and propagating information. When we decide where to place sensors, which algorithm to use for processing the data acquired via these sensors, what is the best way to propagate information (e.g., which teaching techniques is the best for each task)—we rely mostly on our own intuition and on the opinion of the experts, and not—as in many other areas—on computer-based systems.

As a result, the methods of data acquisition, processing, and propagation that we select are often far from being optimal. This non-optimality is not surprising: when a driver navigates in a new town, clearly a GPS-based navigator will select a much better route than a driver would come up based on his or her intuition.

State-of-the-art in modeling and enhancing knowledge-related processes: successes and limitations. To make effective recommendation on how to acquire, process, and propagate knowledge, it is necessary to build realistic models of the corresponding processes, and then use these models to find optimal ways of controlling these processes.

This need is well understood. There are many numerical models of knowledge acquisition, processing, and propagations; see, e.g., [1–12]. Some of these models have been successfully used to enhance the corresponding processes.

L.O. Lerma and V. Kreinovich, *Towards Analytical Techniques for Optimizing Knowledge Acquisition, Processing, Propagation, and Use in Cyberinfrastructure and Big Data*, Studies in Big Data 29, DOI 10.1007/978-3-319-61349-9_1

However, these applications are limited by the fact that most of these models are based on detailed numerical simulation of the corresponding processes. Knowledge-related processes are complex, and therefore, the existing simulations are very complex—and optimizations based on these models are also very complex and time-consuming. These models have been successfully used in large-scale applications: e.g., to optimize how a large research laboratory works—but they are not yet fully ready for everyday applications to day-to-day practical decisions on where to place sensors, which algorithms to select, how to best propagate knowledge—decisions which researchers face all the time.

This is especially important in situations of *big data*, when the amount of data is so huge that the traditional numerical methods are not applicable; see, e.g., [13–16].

Need for analytical models. To help researchers make these decisions, we need to develop not only *numerical* simulation models, we also need to develop *analytical* models for the corresponding knowledge-related processes, models that would allow easier optimization and application.

It is mostly analytical models which have led to successful applications of science and engineering. For example, in chemistry, we can write down Schroedinger's equations which simulate how electrons, atoms, and molecules interact—but for complex molecules, this requires large computations on high performance computers. In most practical applications, chemists use simplified analytical models to predict and control the results of chemical reactions. Similarly, in statistical physics, we can write Newton's equations for the motion of all the molecules—and sometimes it is necessary—but most applications use analytical macromodels operating with temperature, pressure, and other macro-characteristics instead of an atom-by-atom descriptions.

What we do in this book: main objective and chapter-by-chapter structure. The main purpose of this book is to develop analytical models for all the knowledge-related processes, from knowledge acquisition to knowledge processing and knowledge propagation. Of course, this is vast area of research, and we are not aiming at covering all possible aspects which can be described by analytical models. Our goal is more practical: to develop analytical models for at least some of the knowledge-related processes related to specific interdisciplinary cyberinfrastructure-related research projects (some of which involve processing big data). In this book, we present the results of our work.

In Chap. 2, we start with data acquisition. The main source of knowledge is processing data. Data comes from sensors. Within a limited budget, it is extremely important to make sure that the use of the sensors is optimized so that we get the largest possible amount of useful data from these sensors.

Traditionally, most data comes from stationary sensors, i.e., sensors which we place at fixed locations. For such sensors, it is important to come up with the optimal placement, the placement which would lead to the largest amount of useful data. We analyze this problem in Sect. 2.1, on the example of placing bio-weapon detectors, and in Sect. 2.2, on the example of placing environmental sensors.

The problem of optimal use becomes more technically challenging if we take into account the possibility of using mobile sensors, i.e., sensors which we can move along different trajectories. In this case, it is important to come up with optimal trajectories, i.e., the trajectories which would lead to the largest amount of useful data. We analyze this problem in Sect. 2.3, on the example of Unmanned Aerial Vehicles (UAVs) patrolling the border.

In all these cases, it is important to make sure that not only we have an algorithm producing the optimal placement or optimal trajectory: we also need to make sure that the corresponding algorithms are computationally efficient, i.e., that the corresponding optimization algorithms can produce the resulting optimal setting in reasonable time. The more sensors we need to place, the more computations we need and therefore, the more important it is for the computation time to be reasonable. We analyze this problem in Sect. 2.4, again on the example of security problems.

Once the data is collected, we need to process this data. Problems related to data processing are analyzed in Chap. 3. For processing, we need to use computers—and the more data we collect, the more computer power we need. It is therefore important to optimally distribute this computing power. This is the problem that we will analyze in Sect. 3.1.

In many cases, data processing is a creative process, it goes beyond simple application of known algorithms. The problem of selecting the best ways of organizing this process is handled in Sect. 3.2.

To come up with the best ways of processing data, of extracting knowledge from the data, it is often not sufficient to have *individual* efforts, we need creative *teams*—teams that combine domain expertise and computer expertise. Just like with computers, simply bringing people together does not always improve their efficiency. It is therefore important to make sure that people collaborate in the most efficient way. This aspect of data and knowledge processing is analyzed in Sect. 3.3.

Once we have transformed data into knowledge, we need to propagate this knowledge—so that other researchers can use and enhance this knowledge. Problems related to data propagation are analyzed in Chap. 4. To propagate data, first, we need to motivate people to learn the new knowledge, we need to make sure that the idea is propagated to more and more people. To ensure that, we need to analyze the process of idea propagation; this is done in Sect. 4.1.

Once a person is willing to learn the corresponding techniques and ideas, we can start the actual learning. For this learning to be successful, we need to get a good understanding of where the person stands now, what is his/her level of knowledge in the corresponding areas. This assessment problem is analyzed in Sect. 4.2.

Once this information is known, we need to actually present this information to the interested folks—and use appropriate feedback to modify (if needed) the speed with which this knowledge is presented. Issues related to the material's presentation are analyzed in Sects. 4.3 and 4.4. Specifically, in Sect. 4.3, we consider the problem from the global viewpoint: e.g., in what order we should present different parts of the material, and how much flexibility should we give to students. In Sect. 4.4, we consider this problem from the local viewpoint: what is the best way to present different items. Finally, in Sect. 4.5, we analyze the problems related to feedback.

In Chap. 5, we analyze the problems related to using data. How can we use the acquired knowledge? In many practical situations, we have a well-defined problem, with a clear well-formulated objective. Such problems are typical in engineering: we want a bridge which can withstand a given load, we want a car with a given fuel efficiency, etc. There exist many techniques for solving such well-defined optimization problems.

However, in many practical situations, we only have partial and/or subjective information about the situation and about our objectives. In such situations, we need to make decisions under uncertainty. This aspect of knowledge use is what we analyze in Chap. 5.

The ultimate goal of knowledge use is to help the users. To do this, we need to get a good understanding of the corresponding processes. Gaining such an understanding is the main objective of *science*, when we use the observed data to find the dependencies between different quantities. Once these dependencies have been discovered, we can apply this knowledge to help the users: we can find out how to better *control* the existing systems, we can find out how to better *design* the new systems, and we can find out how to better *maintain* the systems. In all these engineering tasks, we are interested in decision making under uncertainty, in particular, in taking imprecise expert knowledge into account.

In Chap. 5, we provide examples of applications to science and to all three aspects of engineering. Specifically, in Sect. 5.1, we consider an example of an application to science, in Sect. 5.2, we consider an example of an application to control, in Sect. 5.3, we deal with an application to design, and in Sect. 5.4, we consider an application to maintenance.

Finally, Chap. 6 contains conclusions.

References

1. G. Aletti, A.K. Naimzada, G. Naldi, Mathematics and physics applications in sociodynamics simulation: the case of opinion formation and diffusion, in *Mathematical Modeling of Collective Behavior in Socio-Economic and Life Sciences*, ed. by G. Naldi, L. Pareschi, G. Toscani (Springer, Dordrecht, 2010), pp. 203–221
2. L.M.A. Bettencourt, A. Cintron-Arias, D.I. Kaiser, C. Castillo-Chavez, The power of a good idea: quantitative modeling of the spread of ideas from epidemiological models. Phys. A **364**, 513–536 (2006)
3. L.M.A. Bettencourt, D.I. Kaiser, J. Kaur, C. Castillo-Chavez, D.E. Wojick, Population modeling of the emergence and development of scientific fields. Scientometr. **75**(3), 495–518 (2008)
4. P.A. Geroski, Models of technology diffusion. Res. Policy **29**, 603–625 (2000)
5. M. Gomez-Rodriguez, J. Leskovec, A. Krause, Inferring networks of diffusion and influence, in *Proceedings of the 16th ACM SIGKDD International Conference on Knowledge Discovery and Data Mining KDD'10*, Washington, DC, 25–28 July 2010
6. M. Ingawale, Understanding the Wikipedia phenomenon: a case for agent based modeling, in *Proceedings of the ACM Ph.D. Workshop on Information and Knowledge Management PIKM'08*, Napa Valley, California, 30 Oct 2008, pp. 81–84

7. P. Morone, R. Taylor, *Knowledge Diffusion and Innovation: Modeling Complex Entrepreneurial Behaviors* (Edward Elgar Publication, Cheltenham, UK, 2010)
8. A. Pentland, *Social Physics: How Good Ideas Spread—The Lessons from a New Science* (Penguin Press, New York, 2014)
9. G. Rossman, *Climbing the Charts: What Radio Airplay Tells Us about the Diffusion of Information* (Princeton University Press, Princeton, New Jersey, 2012)
10. C.M. Tsai, The knowledge diffusion model associated with innovative knowledge. Expert Syst. Appl. **36**, 11323–11331 (2009)
11. J. Xu, M. Quaddus, Examining a model of knowledge management systems adoption and diffusion: a partial Least Square approach. Knowl. Based Syst. **27**, 18–28 (2012)
12. P. Yan, X. Yang, To trade or to teach: modeling tacit knowledge diffusion in complex social networks, in *Proceedings of the Second International Conference on Future Information Technology and Management Engineering FITME'2009*, Sanya, China, 13–14 Dec 2009, pp. 151–154
13. B. Baesens, *Analytics in a Big Data World: The Essential Guide to Data Science and its Applications* (Wiley, Hoboken, New Jersey, 2014)
14. B. Marr, *Big Data: Using SMART Big Data, Analytics and Metrics to Make Better Decisions and Improve Performance* (Wiley, Chichester, UK, 2015)
15. N. Marz, J. Warren, *Big Data: Principles and Best Practices of Scalable Realtime Data Systems* (Manning Publication, Shelter Island, New York, 2015)
16. V. Mayer-Schönberger, K. Cukier, *Big Data: A Revolution That Will Transform How We Live, Work, and Think* (John Murray Publication, London, UK, 2013)

Chapter 2
Data Acquisition: Towards Optimal Use of Sensors

Abstract The main source of knowledge is processing data. Data comes from sensors. Within a limited budget, it is extremely important to make sure that the use of the sensors is optimized so that we get the largest possible amount of useful data from these sensors. Traditionally, most data comes from stationary sensors, i.e., sensors which we place at fixed locations. For such sensors, it is important to come up with the optimal placement, the placement which would lead to the largest amount of useful data. We analyze this problem in Sect. 2.1, on the example of placing bio-weapon detectors, and in Sect. 2.2, on the example of placing environmental sensors. The problem of optimal use becomes more technically challenging if we take into account the possibility of using mobile sensors, i.e., sensors which we can move along different trajectories. In this case, it is important to come up with optimal trajectories, i.e., the trajectories which would lead to the largest amount of useful data. We analyze this problem in Sect. 2.3, on the example of Unmanned Aerial Vehicles (UAVs) patrolling the border. In all these cases, it is important to make sure that not only we have an algorithm producing the optimal placement or optimal trajectory: we also need to make sure that the corresponding algorithms are computationally efficient, i.e., that the corresponding optimization algorithms can produce the resulting optimal setting in reasonable time. The more sensors we need to place, the more computations we need and therefore, the more important it is for the computation time to be reasonable. This is especially important in situations of *big data*, when the amount of data is so huge that the traditional numerical methods are not applicable [1–4]. We analyze this problem in Sect. 2.4, again on the example of security problems.

2.1 Optimal Use of Stationary Sensors: Case Study of Optimal Placement of Bio-Weapon Detectors

In this section, we analyze the problem of the optimal use of stationary sensors, on the example of optimal placement of bio-weapon detectors. Biological weapons are difficult and expensive to detect. Within a limited budget, we can afford a limited number of bio-weapon detector stations. It is therefore important to find the optimal

L.O. Lerma and V. Kreinovich, *Towards Analytical Techniques for Optimizing Knowledge Acquisition, Processing, Propagation, and Use in Cyberinfrastructure and Big Data*, Studies in Big Data 29, DOI 10.1007/978-3-319-61349-9_2

locations for such stations. A natural idea is to place more detectors in the areas with more population—and fewer in desert areas, with fewer people. However, such a commonsense analysis does not tell us how many detectors to place where. To decide on the exact placement of bio-weapon detectors, we formulate the placement problem in precise terms, and come up with an (almost) explicit solution to the resulting optimization problem.

The results from this section were first published in [5].

Formulation of the practical problem. Biological weapons are difficult and expensive to detect. Within a limited budget, we can afford a limited number of bio-weapon detector stations. It is therefore important to find the optimal locations for such stations.

Commonsense analysis of the problem. A natural idea is to place more detectors in the areas with more population—and fewer in areas with fewer people, e.g., in the desert areas. However, such a commonsense analysis does not tell us how many detectors to place where. To decide on the exact placement of bio-weapon detectors, we must formulate the placement problem in precise terms.

Objective function. The above commonsense idea is based on a (reasonable) assumption that the adversary's objective is to kill as many people as possible. Vice versa, our objective is to minimize the potential effect of a bio-weapon attack.

Comment. In this chapter, we mainly concentrate on the above objective function. This objective function may not always fully describe the adversary's objectives. For example, one of the objectives of political terrorism may be extra publicity for the cause. From this viewpoint, an adversary may prefer a scenario with a smaller number of victims if several of these victims are well-known. It is therefore desirable to formulate the objective functions that describe this (and similar) approaches, and extend our optimization analysis to the case of such more complex objective functions.

Towards precise formulation of the problem: what is known. Since the objective is to target as many people as possible, to analyze this situation, we need to know how many people live at different locations. In precise terms, we assume that we know, for every possible location x, the population density $\rho(x)$ in the vicinity of this location.

We assume that we know the number N of detectors that we can afford to place in the given territory.

We also assume that we know the efficiency of a bio-weapons detector station. We will estimate this efficiency by the distance d_0 at which this station can detect an outbreak of a disease.

For many diseases, $d_0 = 0$—we can only detect a disease when the sources of this disease reach the detecting station.

However, it is quite possible that for some diseases, we have a super-sensitive equipment that is able to detect the concentration of the bio-weapons agent at a level below the threshold that makes this agent dangerous to the population. In this case, we can detect the coming disease before it starts affecting people in the direct vicinity of the station—i.e., in effect, we have $d_0 > 0$.

For simplicity, we assume that the disease spreads equally fast in all directions.

Comment. This is also a somewhat simplifying assumption, since in reality, a disease spreads

- either with human movements—in which case in the vicinity of an interstate it spreads faster in the direction of the interstate,
- or with wind—in which case it spreads faster in the direction of the prevailing winds.

How we can describe the detector placement. On a large-scale basis, we need to decide how many detectors to place in different areas. In other words, we need to find the *density* $\rho_d(x)$ of detector placement—the number of detectors per unit of area (e.g., a square mile).

Under this description, the number of detectors in an area of size Δx is approximately equal to $\rho_d(x) \cdot \Delta x$, so the overall number of detectors can be obtained by adding these amounts, as $\int \rho_d(x)\,dx$. Thus, the constraint that we have exactly N detecting stations can be described as

$$\int \rho_d(x)\,dx = N. \tag{2.1.1}$$

Optimal placement of sensors: at the vertices of a hexagonal grid. We want to place the sensors in such a way that the largest distance D to a sensor is as small as possible. Alternatively, if D is fixed, we want to minimize the number of sensors for which every point is at a distance $\le D$ from one of the sensors. In geometric terms, this means that every point on a plane belongs to a circle of radius D centered on one the sensors—and thus, the whole plane is covered by such circles. Out of all such coverings, we want to find the covering with the smallest possible number of sensors.

It is known that the smallest such number is provided by an equilateral triangle grid, i.e., a grid formed by equilateral triangles; see, e.g., [6]. Hence, in this section, we will select such a grid.

Locations of detector stations are assumed to be known to the adversary. Bio-weapon detector stations are not easily concealable. Thus, we assume that the adversary knows the locations of different stations.

How to estimate the effect of placing bio-weapons at a location x. Let us assume that we have already decided how many detectors to place in different regions, i.e., that we have already selected the density function $\rho_d(x)$.

Within a small region of area A, we have $A \cdot \rho_d(x)$ detectors. Thus, if we, e.g., place these detectors on a grid with distance h between the two neighboring ones in each direction, we have:

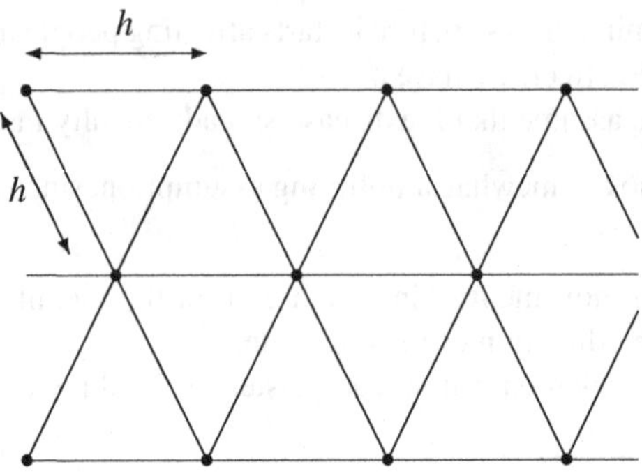

For this placement, the set of all the points which are closest to a given detector forms a hexagonal area:

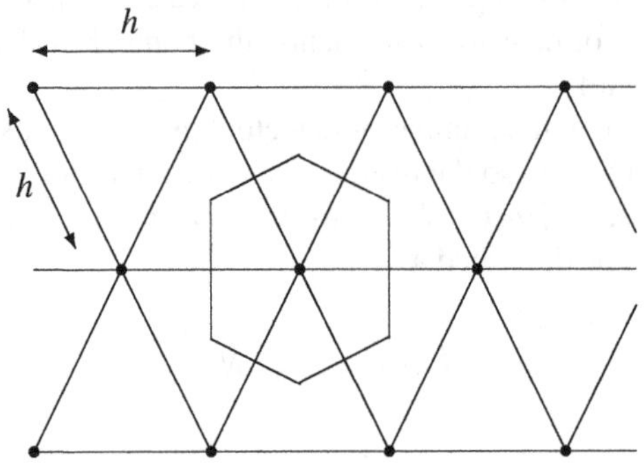

This hexagonal area consists of 6 equilateral triangles with height $h/2$:

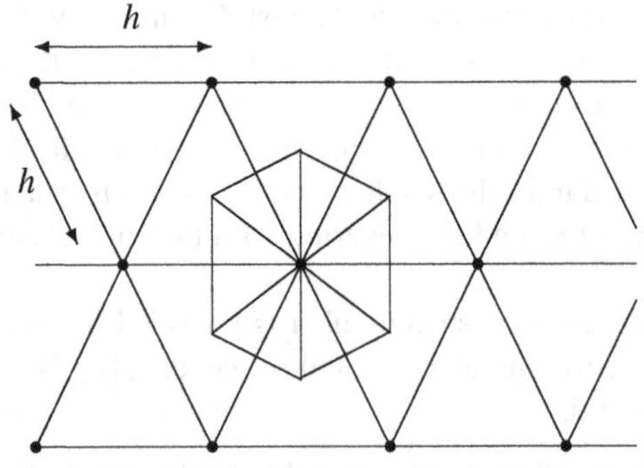

In each triangle, the height $h/2$ is related to the size s by the formula

$$\frac{h}{2} = s \cdot \cos(60^\circ) = s \cdot \frac{\sqrt{3}}{2}, \tag{2.1.2}$$

hence

$$s = \frac{h}{\sqrt{3}} = h \cdot \frac{\sqrt{3}}{3}. \tag{2.1.3}$$

Thus, the area A_t of each triangle is equal to

$$A_t = \frac{1}{2} \cdot s \cdot \frac{h}{2} = \frac{1}{2} \cdot \frac{\sqrt{3}}{3} \cdot \frac{1}{2} \cdot h^2 = \frac{\sqrt{3}}{12} \cdot h^2. \tag{2.1.4}$$

So, the area A_s of the whole set is equal to 6 times the triangle area:

$$A_s = 6 \cdot A_t = \frac{\sqrt{3}}{2} \cdot h^2. \tag{2.1.5}$$

Each point from the region is the closest to one of the points from the detector grid, so the region of area A is thus divided into $A \cdot \rho_d(x)$ (practically) disjoint sets of area $\frac{\sqrt{3}}{2} \cdot h^2$. So, the area of the region is equal to the sum of the areas of these sets:

$$A = (A \cdot \rho_d(x)) \cdot \frac{\sqrt{3}}{2} \cdot h^2. \tag{2.1.6}$$

Dividing both sides of this equality by A, we conclude that

$$1 = \rho_d(x) \cdot \frac{\sqrt{3}}{2} \cdot h^2, \tag{2.1.7}$$

and hence, that

$$h = \frac{c_0}{\sqrt{\rho_d(x)}}, \tag{2.1.8}$$

where we denote

$$c_0 \stackrel{\text{def}}{=} \sqrt{\frac{2}{\sqrt{3}}}. \tag{2.1.9}$$

From the viewpoint of the adversary, it is desirable to place the bio-weapon at a location which is the farthest away from the detectors—so that it will take the longest time to be detected. For the grid placement, this location is at one of the vertices of the hexagonal zone—at which the distance from each neighboring detector is equal to $s = h \cdot \frac{\sqrt{3}}{3}$. By using formula (2.1.8), we can determine s in terms of $\rho_d(x)$, as

$$s = \frac{c_1}{\sqrt{\rho_d(x)}}, \tag{2.1.10}$$

where we denote

$$c_1 = \frac{\sqrt{3}}{3} \cdot c_0 = \frac{\sqrt[4]{3} \cdot \sqrt{2}}{3}. \tag{2.1.11}$$

Once the bio-weapon is placed at this location, it starts spreading until its spread area reaches the threshold distance d_0 from the detector. In other words, it spreads for the distance $s - d_0$. During this spread, the disease covers the circle of radius $s - d_0$ and area $\pi \cdot (s - d_0)^2$.

By using the known population density $\rho(x)$, we can conclude that the number of affected people $n(x)$ is equal to

$$n(x) = \pi \cdot (s - d_0)^2 \cdot \rho(x). \tag{2.1.12}$$

Substituting the expression (2.1.10) into this formula, we conclude that

$$n(x) = \pi \cdot \left(\frac{c_1}{\sqrt{\rho_d(x)}} - d_0 \right)^2 \cdot \rho(x). \tag{2.1.13}$$

Adversary's choice of the location. According to our assumption about the adversary's objective function, the adversary wants to maximize the number of affected people. Thus, the adversary will select a location x for which this number $n(x)$ (as described by the expression (2.1.13)) is the largest possible. The resulting damage n is thus equal to the largest of the values $n(x)$:

$$n = \max_x \left(\pi \cdot \left(\frac{c_1}{\sqrt{\rho_d(x)}} - d_0 \right)^2 \cdot \rho(x) \right). \tag{2.1.14}$$

Our objective. Our objective is to minimize this overall damage, i.e., to select the detector placement $\rho_d(x)$ so as to minimize this value n.

In other words, we want to *minimize* the worst-possible (maximal) damage. This *minimax* formulation is typical for *zero-sum games*, in which the interests of the two sides are exactly opposite; see, e.g., [7].

Thus, we arrive at the following problem:

Resulting formulation of the problem in precise terms. We are given the population density $\rho(x)$, the value d_0, and the total number of detectors N. We want to find a function $\rho_d(x)$ that minimizes the expression (2.1.14) under the constraint $\int \rho_d(x)\,dx = N$.

Analysis of the resulting optimization problem. The damage is determined by the maximum n of the function $n(x)$. Let us assume that we have already selected the optimal detector density function, i.e., the function $\rho_d(x)$ that minimizes the desired objective function n.

Let us show that the damage function $n(x)$ corresponding to this selection is constant. We will prove this by contradiction. If the function $n(x)$ is not constant, this means that at some locations x, the values $n(x)$ are smaller than the maximum n. In this case, we can slightly increase the detector density at the locations where $n(x) = n$, at the expense of slightly decreasing the location density at locations where $n(x) < x$.

The value of the expected damage $n(x)$ monotonically decreases with the detector density $\rho_d(x)$. This mathematical observation is in perfect accordance with common sense: the more detectors we place at some location, the earlier we will be able to detect bio-weapons and thus, the smaller will be the resulting damage.

Thus, the above re-arrangement of detectors will decrease the value of $n(x)$ at all locations where $n(x) = n$—and slightly increase at all other locations. As a result, after this detector relocation, the overall maximum $n = \max_x n(x)$ will decrease. This possibility contradicts to our initial assumption that the value n is the smallest possible. Thus, the function $n(x)$ is indeed constant.

Let us denote this constant by n_0. Then, from the formula (2.1.13) for $n(x)$, we conclude that

$$n_0 = \pi \cdot \left(\frac{c_1}{\sqrt{\rho_d(x)}} - d_0 \right)^2 \cdot \rho(x). \tag{2.1.15}$$

Thus, we conclude that

$$\left(\frac{c_1}{\sqrt{\rho_d(x)}} - d_0 \right)^2 = \frac{n_0}{\pi \cdot \rho(x)}, \tag{2.1.16}$$

$$\frac{c_1}{\sqrt{\rho_d(x)}} - d_0 = \frac{c_2}{\sqrt{\rho(x)}}, \tag{2.1.17}$$

where we denote

$$c_2 \stackrel{\text{def}}{=} \frac{\sqrt{n_0}}{\sqrt{\pi}}. \tag{2.1.18}$$

Thus, we get

$$\frac{c_1}{\sqrt{\rho_d(x)}} = d_0 + \frac{c_2}{\sqrt{\rho(x)}}, \tag{2.1.19}$$

$$\sqrt{\rho_d(x)} = \frac{c_1}{d_0 + \dfrac{c_2}{\sqrt{\rho(x)}}}, \tag{2.1.20}$$

and

$$\rho_d(x) = \frac{c_1^2}{\left(d_0 + \dfrac{c_2}{\sqrt{\rho(x)}} \right)^2}, \tag{2.1.21}$$

From Eq. (2.1.11), we conclude that

$$c_1^2 = \frac{2 \cdot \sqrt{3}}{9}, \tag{2.1.22}$$

hence

$$\rho_d(x) = \frac{2\cdot\sqrt{3}}{9}\cdot\frac{1}{\left(d_0 + \frac{c_2}{\sqrt{\rho(x)}}\right)^2}. \tag{2.1.23}$$

The value c_2 must be determined from the equation (2.1.1).

Thus, we arrive at the following solution:

Solution: The optimal detector location is characterized by the detector density

$$\rho_d(x) = \frac{2\cdot\sqrt{3}}{9}\cdot\frac{1}{\left(d_0 + \frac{c_2}{\sqrt{\rho(x)}}\right)^2},$$

where the parameter c_2 must be determined from the equation

$$\int \frac{2\cdot\sqrt{3}}{9}\cdot\frac{1}{\left(d_0 + \frac{c_2}{\sqrt{\rho(x)}}\right)^2}\,dx = N. \tag{2.1.24}$$

Case of $d_0 = 0$. As we have mentioned earlier, in some cases, we have $d_0 = 0$. In this case, the formula (2.1.23) takes a simplified form

$$\rho_d(x) = C\cdot\rho(x) \tag{2.1.25}$$

for some constant C. In this case, the detector density is exactly proportional to the population density.

Substituting the expression (2.1.25) into the constraint (2.1.1), we conclude that

$$N = C\cdot N_p, \tag{2.1.26}$$

where $N_p = \int \rho(x)\,dx$ is the total population. Thus, $C = \frac{N}{N_p}$ and the optimal detector placement Eq. (2.1.25) takes the form

$$\rho_d(x) = \frac{N}{N_p}\cdot\rho(x). \tag{2.1.27}$$

Towards more relevant objective functions. In our computations, we assumed that the main objective of the adversary is to maximize the number of people affected by the bio-weapon, i.e., to maximize the value $\int_A \rho(x)\,dx$, where A is the region were people become affected before the bio-weapon is detected.

As we have mentioned, the actual adversary's objective function may differ from this simplified objective function. For example, the adversary may take into account that different locations have different publicity potential. In this case, instead of maximizing the total number of affected people, the adversary may want to maximize the weighted value $\int_A \widetilde{\rho}(x)\,dx$, where $\widetilde{\rho}(x) \stackrel{\text{def}}{=} w(x) \cdot \rho(x)$, and the weight $w(x)$ describes the publicity-related importance of the location x.

From the purely mathematical viewpoint, once we have fixed the weight functions $w(x)$, we get the exact same problem as before—with the only difference that we now have "effective population density" $\widetilde{\rho}(x)$ instead of the original density $\rho(x)$. Thus, if we know the exact weight function $w(x)$, then we find the optimal detector density $\rho_d(x)$ by substituting the effective population density $\widetilde{\rho}(x)$ instead of $\rho(x)$ into the above formulas.

2.2 Optimal Use of Stationary Sensors: Case Study of Optimal Placement of Environmental Sensors

In this section, we analyze the problem of the optimal use of stationary sensors, on another example: of optimal placement of environmental sensors. Specifically, we show that under reasonable assumption, the spatial variability of a field $f(x)$, i.e., the expected value

$$F(z) \stackrel{\text{def}}{=} E[(f(x+z) - f(x))^2],$$

has the form $F(z) = \left|\sum_{i=1}^{n}\sum_{j=1}^{n} g_{ij} \cdot z_i \cdot z_j\right|^{\alpha}$. We explain how to find g_{ij} and α from the observations, and how to optimally place sensors in view of this spatial variability.

The results of this section were first published in [8].

Need to describe spatial variability. To understand climate trends, we need to describe not only the values of temperature, humidity, wind speed and direction at a single location, we also need to know how these characteristics change from one location to the other. In other words, we need to describe spatial variability of the corresponding characteristics.

There is a similar need in other application areas. For example, to understand the brain activity within a region, in addition to describing brain activity at certain locations, we also need to describe how this brain activity changes from one location to the other, i.e., we need to describe spatial variability of the corresponding characteristics.

How to describe spatial variability: use of random variables. In general, we have a characteristic $f(x)$ that takes different values at different locations x. Since we cannot exactly predict the exact future values $f(x)$, it is reasonable to consider them random variables. Random variables $f(x)$ corresponding to different locations x form a *random field*.

How to describe spatial variability: use of normal distributions. The values $f(x)$ are determined by a large number of different factors. In statistics, the joint effect of many small independent factors is—due to the Central Limit Theorem—well described by a normal distribution; see, e.g., [9]. Thus, it is reasonable to assume that the variables $f(x)$ are normally distributed.

A normal distribution is uniquely determined by its first two moments, i.e., by the expected values $E[f(x)]$ and $E[f(x) \cdot f(y)]$. The values $E[f(x)]$ and $E[(f(x))^2]$ describe the behavior at a single location. Thus, to describe spatial variability, it is sufficient to describe the values $E[f(x) \cdot f(y)]$ for $x \neq y$. Since we know the values $E[(f(x))^2]$ and $E[(f(y))^2]$, describing $E[f(x) \cdot f(y)]$ is equivalent to describing the following expected value:

$$C(x, y) \stackrel{\text{def}}{=} E[(f(y) - f(x))^2] = E[(f(y))^2] + E[(f(x))^2] - 2E[f(x) \cdot f(y)].$$

Homogeneity. Locally, the distribution is usually homogenous, i.e., does not change after a shift. Thus, if we change x to $x+z$ and y to $y+z$, we should get the same value $C(x, y)$: $C(x+z, y+z) = C(x, y)$. For $z = -z$, this leads to $C(x, y) = C(0, y-x)$. So, to describe spatial variability, it is sufficient to describe the function

$$F(z) \stackrel{\text{def}}{=} C(0, z) = E[(f(x+z) - f(x))^2].$$

Comment. For $z = 0$, the above definition leads to $F(0) = 0$.

Other natural requirements. It is reasonable to assume that $F(z)$ continuously depends on z.

It is also reasonable to assume that there is spatial variability, i.e., that $F(z) > 0$ for $z > 0$.

Another requirement is that $f(x)$ is very close to $f(y)$ only for close x and y. Formally, we will require that for some value $F_0 > 0$, the set $\{z : F(z) \leq F_0\}$ is bounded.

Comment. It should be mentioned that the spatial distribution is often *anisotropic*, i.e., depends on the direction. For example, a North-South oriented mountain range goes through the city of El Paso. The closeness to the mountain affects temperature, rainfall, wind, etc. As a result, the meteorological characteristics change much more when we move in the East-West direction than when we move in the North-South one.

We need to select a few-parametric family of functions $F(z)$. In different practical situations, we have different functions $F(z) \geq 0$. To describe all such situations, it is desirable to have a parametric family $\mathscr{F}$ of possible functions $F(z)$.

Often, we only have a limited amount of data, so we can only statistically significantly determine a small number of parameters of the function $F(z)$. For example, in environmental sciences, we have a limited number of observations in remote areas such as most areas of Arctic and Antarctica. In brain research, we also often only have limited data. To cover such situations, it is desirable to have simple, few-parametric families $\mathcal{F}$.

Desired properties of few-parametric families. The numerical value of a physical characteristic depends on the choice of a measuring unit. For example, for length, if we change from inches to cm, the numerical values increase by 2.54. In general, if we use a new unit which is λ times smaller than the previous one, then numerical values $f(x)$ increase by λ, and the resulting values of $F(z)$ increase by λ^2. In principle, we can have an arbitrary positive value $C = \lambda^2$, so it is reasonable to require that the family $\mathcal{F}$ contains, with every function $F(z)$, also all functions $C \cdot F(z)$ for every $C > 0$.

Another possible change is a change in spatial coordinates. In some applications, the usual coordinates work best, in other applications, polar, cylindrical, or other coordinates are more appropriate. Locally, each smooth coordinate transformation $x_i \to f_i(x_1, \ldots, x_n)$ can be well approximated by a linear function $x_i \to \sum\limits_{j=1}^{n} a_{ij} \cdot x_j + a_i$, i.e., in matrix terms, $x \to Ax + a$. Under this transformation, the difference $z = y - x$ is replaced with Az. It is therefore reasonable to require that the the family $\mathcal{F}$ contains, with every function $F(z)$, also all functions $F(Az)$ for all non-degenerate matrices A.

It turns out that these two requirements are sufficient to determine few-parametric families $\mathcal{F}$ with the smallest possible number of parameters.

Proposition. *Let $\mathcal{F}$ be a $\dfrac{n \cdot (n+1)}{2}$-parametric family of continuous functions $F(z)$ from $\mathbb{R}^n$ to $\mathbb{R}$ for which $F(z) = 0$, $F(z) > 0$ for $z \neq 0$, and for some $F_0 > 0$, the set $\{z : F(z) \leq F_0\}$ is bounded. Let us also assume that the family $\mathcal{F}$ contains, with every function $F(z)$, also all functions $C \cdot F(z)$ for all $C > 0$ and all functions $F(Az)$ for all non-degenerate matrices A. Then, every function $F \in \mathcal{F}$ has the form*

$$F(z) = \left| \sum_{i=1}^{n} \sum_{j=1}^{n} g_{ij} \cdot z_i \cdot z_j \right|^{\alpha}$$

for some real values α and g_{ij}.

Proof. In this proof, similarly to [10], we will use ellipsoids centered at 0, i.e., ellipsoids $E = \{z : \sum g_{ij} \cdot z_i \cdot z_j \leq 1\}$. We will call them *c-ellipsoids* (c for *centered*). To describe all such c-ellipsoids, we need to describe all symmetric matrices g_{ij}, so the family of c-ellipsoids is $\dfrac{n \cdot (n+1)}{2}$-dimensional. The border $\{z : \sum g_{ij} \cdot z_i \cdot z_j = 1\}$ of an ellipsoid E will be denoted by ∂E.

1°. Let $F \in \mathcal{F}$. Let us first prove that there is a c-ellipsoid E_0 on whose border ∂E_0 we have $F(z) = F_0$ for all $z \in \partial E_0$.

1.1°. By definition of the class $\mathcal{F}$, the set $S \stackrel{\text{def}}{=} \{z : F(z) \leq F_0\}$ is bounded, and each function $F \in \mathcal{F}$ is continuous. Since $F(z)$ is continuous, the set S is closed.

Every bounded set can be enclosed into a c-ellipsoid. It is known (see, e.g., [11]) that, among all ellipsoids containing a given closed bounded set, there is exactly one ellipsoid with the smallest volume.

Let E_0 denote the c-ellipsoid with the smallest volume that contains the set S. We will say that this ellipsoid *corresponds* to the function $F(z)$.

Comment. The existence of the smallest-volume ellipsoid follows from the fact that every continuous function on a compact set attains its minimum. Uniqueness follows from the fact that if we have two c-ellipsoids E and E' of the same volume containing the same set, then we can select coordinates in which both matrices are diagonal, i.e., have the form $\sum g_i \cdot z_i^2 \le 1$ and $\sum g_i' \cdot z_i^2 \le 1$; then, for $g_i'' = \dfrac{g_i + g_i'}{2}$, the ellipsoid $\sum g_i'' \cdot z_i^2 \le 1$ also contains the bounded set and, as can be easily shown, has a strictly smaller volume than E and E'.

1.2°. It is known that every c-ellipsoid E in appropriate affine coordinates becomes a unit ball $\{z : \sum z_i^2 \le 1\}$. In other words, every ellipsoid can be obtained from a unit ball by an appropriate affine transformation. By combining the affine transformations corresponding to E and to E_0, we conclude that E can be obtained from the ellipsoid E_0 by an affine transformation $z \to Az$.

Under an affine transformation, the ratio of volumes is preserved. So, since E_0 the c-ellipsoid with the smallest volume contains the set $S = \{z : F(z) \le F_0\}$, E is the c-ellipsoid with the smallest volume containing the set $S' = \{z : F'(z) \le F_0\}$, where $F'(z) \stackrel{\text{def}}{=} F(Az) \in \mathcal{F}$.

Different ellipsoids correspond to different functions $F'(z)$, so we have as many such functions $F'(z)$ as there are ellipsoids—i.e., a $\dfrac{n \cdot (n+1)}{2}$-dimensional family.

1.3°. There are many affine transformations (rotations) that preserve the ball; in particular, for every two points on a unit sphere, there is a rotation that transforms one into another.

Thus, there are many affine transformations that preserve every ellipsoid E. In particular, for every two points $z, z' \in \partial E$ on this ellipsoid's border, there is an affine transformation that preserves ∂E and transforms z into z'.

For the ellipsoid E_0, let us denote, by G_0, the group of all affine transformations that preserve ∂E_0.

1.4°. Let us show that the border ∂E_0 of the ellipsoid E_0 contains some points from the set $S \stackrel{\text{def}}{=} \{z : F(z) \le z_0\}$.

We will prove this by contradiction. Let us assume that the border ∂E_0 of the ellipsoid E_0 does not contain any points from the set S. Then, we can proportionally shrink E_0 and get a new c-ellipsoid with the smaller volume that still contains S. This contradicts to the fact that E_0 has the smallest volume. The statement is proven.

1.5°. Let us prove that for all $z \in \partial E_0 \cap S$, we have $F(z) = F_0$.

Indeed, since $z \in S$, by definition of the set S, we have $F(z) \le F_0$. On the other hand, since z belongs to the border ∂E_0, the point z is a limit of points z_n from outside E_0: $z_n \to z$. Outside E_0, there are no points from S, so for all $z_n \notin E_0$, we

have $F(z_n) > F_0$. Since the function $F(z)$ is continuous, in the limit $z_n \to z$, we get $F(z) \geq F_0$. From $F(z) \leq F_0$ and $F(z) \geq F_0$, we conclude that $F(z) = F_0$.
1.6°. Finally, let us prove that every point $z \in \partial E_0$ belongs to the set S; due to Part 1.5 of this proof, this will imply that $F(z) = F_0$ for all $z \in \partial E_0$.

We will prove this statement by contradiction. Let us assume that not every point $z \in \partial E_0$ belongs to the set S. Since transformations from G_0 transform every point $z \in \partial E_0$ into every other point $z' \in \partial E_0$, this means that not all transformations from G_0 preserve the intersection $\partial E_0 \cap S$. Thus, transformations that preserve the intersection form a subgroup $G_0' \subset G_0$. Subgroups of the group of rotations are well known, they have smaller dimension than G_0. Thus, we have a finite-parametric family of transformations (of dimension $\geq$1) that preserve ∂E_0 and turn the set $S = \{z : F(z) \leq F_0\}$ into a different set S'—i.e., which turn $F(z)$ into a different function $F'(z)$ for which the ellipsoid E_0 is the same. Thus, we have an at least 1-dimensional family of functions $F'(z)$ corresponding to E_0.

By applying an affine transformation, we get a similar family of functions for every ellipsoid. The family of ellipsoids is already $\frac{n \cdot (n+1)}{2}$-dimensional, and for each of them, there is an $\geq$1-dimensional family of functions—thus, we get a $\geq \left(\frac{n \cdot (n+1)}{2} + 1\right)$-dimensional family of functions $F'(z)$—which contradicts to our assumption that the whole family $\mathscr{F}$ is no more than $\frac{n \cdot (n+1)}{2}$-dimensional. This contradiction shows that indeed $\partial E \subseteq S$.
2°. The ellipsoid E_0 corresponding to the function $F(z)$ has the form $\{z : \|z\|^2 \leq 1\}$, where $\|z\|^2 \stackrel{\text{def}}{=} \sum\limits_{i,j} g_{ij} \cdot z_i \cdot z_j$. Let us prove that the function $F(z)$ has the form $F(z) = h(\|z\|)$ for some function $h(t)$ from real numbers to real numbers.

In other words, we need to prove that for every value v, the function $F(z)$ has a constant value on the border $\partial E_v \stackrel{\text{def}}{=} \{z : \|z\|^2 = v\}$ of the ellipsoid $E_v \stackrel{\text{def}}{=} \{z : \|z\|^2 \leq v\}$ which is obtained from E_0 by an appropriate dilation (homothety).

Indeed, if the function $F(z)$ had two different values on different points $z, z' \in \partial E_v$, then, similarly to Part 1.6 of this proof, we would be able to apply appropriate affine transformations and get a $\geq$1-parametric family of functions $F'(z)$ corresponding to the same ellipsoid E_0 and thus, a $\geq \left(\frac{n \cdot (n+1)}{2} + 1\right)$-dimensional family of functions $F'(z)$—which contradicts to our assumption that $\dim(\mathscr{F}) \leq \frac{n \cdot (n+1)}{2}$.
3°. To complete the proof, let us show that $h(t) = \text{const} \cdot t^\alpha$.

Let us consider the functions $F(z)$ corresponding to all c-ellipsoids E which have the same volume $V(E)$ as E_0: $V(E) = V(E_0)$. The dimension of the family of all such ellipsoids is $\frac{n \cdot (n+1)}{2} - 1$.

For every function $F(z) = h(\|z\|) \in \mathscr{F}$, and for every two real numbers $C > 0$ and $k > 0$, the family $\mathscr{F}$ contains the function $C \cdot F(k \cdot z) = C \cdot h(k \cdot \|z\|)$. The corresponding transformations form a 2-dimensional multiplicative group.

The resulting family of functions cannot be fully 2-dimensional, since then, by considering such a family for every ellipsoid E with $V(E) = V(E_0)$, we would have a family of dimension

$$\geq \left(\frac{n \cdot (n+1)}{2} - 1\right) + 2 = \frac{n \cdot (n+1)}{2} + 1 > \frac{n \cdot (n+1)}{2}$$

inside the family $\mathscr{F}$. Thus, in the 2-dimensional transformation group, there is a ≥ 1-dimensional subgroup that keeps the function $h(t)$ invariant.

All subgroups of the 2-dimensional transformation group are well known, so we have $C(k) \cdot h(k \cdot t) = h(t)$ for some $C(k)$, and hence, $h(k \cdot t) = C^{-1}(k) \cdot h(t)$. It is known (see, e.g., [12]), that every continuous function that satisfies this functional equation has the form $h(t) = A \cdot t^{\alpha}$ for some A and α. The statement is proven, and so is our main result.

Mathematical comment: relation to Riemannian geometry. In general, the values g_{ij} describing spatial variability differ from one location to another. Thus, to describe spatial variability, we need to describe the values $g_{ij}(x)$ corresponding to different locations x. Mathematically, this is equivalent to describing a Riemannian metric.

How to determine g_{ij} and α from the empirical data? Based on the recorded values $f(x, t)$ at different locations x at different times $t = 1, \ldots, T$, we can estimate $C(z) = E[(f(x+z) - f(x))^2]$ as

$$C(z) = \frac{1}{T} \cdot \sum_{t=1}^{T} (f(x+z, t) - f(x, t))^2.$$

We can then use the following iterative procedure to find g_{ij} and α. Initially, we take $g_{ij}^{(0)} = \delta_{ij}$, i.e., $g_{ii}^{(0)} = 1$ and $g_{ij}^{(0)} = 0$ when $i \neq j$. At each iteration k, we start with the values $g_{ij}^{(k-1)}$, and do the following.

First, we estimate $\alpha^{(k)}$ from the condition $C(z) \approx \left|\sum g_{ij}^{(k-1)} \cdot z_i \cdot z_j\right|^{\alpha}$ We can find this α by taking the logarithms of both sides and applying the Least Squares Method to the resulting system of linear equations with unknown α:

$$\ln C(z) \approx \alpha \cdot \ln\left(\sum_{i=1}^{n} \sum_{j=1}^{n} g_{ij}^{(k-1)} \cdot z_i \cdot z_j\right).$$

Once $\alpha^{(k)}$ is computed, we estimate $g_{ij}^{(k)}$ by applying the Least Squares Method to the following system of linear equations with unknown g_{ij}:

$$(C(z))^{1/\alpha^{(k)}} \approx \sum_{i=1}^{n} \sum_{j=1}^{n} g_{ij} \cdot z_i \cdot z_j.$$

Towards optimal sensor location. We want to place the sensors so as to reconstruct the value of $f(x)$ at all locations x with the desired accuracy ε. (Thus, in the spatial direction along which $f(x)$ changes faster we should place sensors more frequently.)

In precise terms, we want to place sensors in such a way that for each spatial location x, there is a sensor location s for which

$$E[(f(x) - f(s))^2 = \left|\sum_{i=1}^{n}\sum_{j=1}^{n} g_{ij} \cdot (x_i - s_i) \cdot (x_j - s_j)\right|^{\alpha} \le \varepsilon^2.$$

For every symmetric matrix g_{ij}, there are affine coordinates—formed by its eigenvectors—in which this matrix become a unit matrix. In this case, the above condition simply means that every location must be ε-close to a sensor location. We have already mentioned, in Sect. 2.1, that under such condition, the asymptotically smallest number of sensors is provided by an equilateral triangle grid, i.e., a grid formed by equilateral triangles [6].

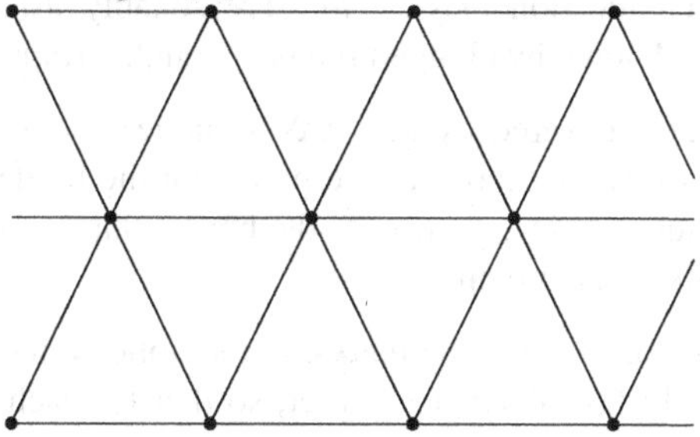

Hence, in general, the sensor grid can be obtained from the equilateral triangle one by an appropriate affine transformation.

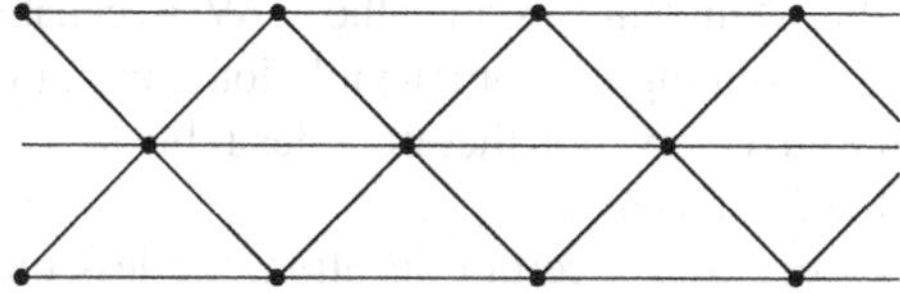

In other words, we should place sensors along the grid parallel to eigenvectors of the matrix g_{ij}.

Mathematical comment: a similar problem of spatial distribution. Instead of spatial *variation*, we can consider a similar problem of spatial *distributions*, i.e., the problem of describing low-dimensional affine-invariant families of probability density functions—families that contain, with every function $\rho(x)$, the function $(\det(A))^{-1} \cdot \rho(Ax + a)$. Similar ellipsoid arguments—but with general ellipsoids instead of c-ellipsoids—show that in this case, every distribution from the corresponding family has the form $\rho(x) = h(\|x - a\|)$ for some function $h(t)$ and some vector a, where $\|z\|^2 = \sum_{i=1}^{n}\sum_{j=1}^{n} g_{ij} \cdot z_i \cdot z_j$ for some values g_{ij}.

2.3 Optimal Use of Mobile Sensors: Case Study of Unmanned Aerial Vehicles Patrolling the Border

In this section, we analyze the problem of the optimal use of mobile sensors, on the example of Unmanned Aerial Vehicles (UAVs) patrolling the border.

The results from this section were first published in [5]

Patrolling the border: a practical problem. Remote areas of international borders can be (and are) used by the adversaries: to smuggle drugs, to bring in weapons. It is therefore desirable to patrol the border, to minimize such actions.

Even with the current increase in the number of border patrol agents, it is not possible to effectively man every single segment of the border. It is therefore necessary to rely on other types of surveillance.

Unmanned Aerial Vehicles (UAVs) are an efficient way of patrolling the border:

- from every location along the border, they provide an overview of a large area, and
- if needed at a different location, they can move reasonably fast to the new location, without being slowed down by clogged roads or rough terrain.

However, while the area covered by the UAV is large, it is still limited. Due to resource limitations, we cannot have all the points on the border under a constant UAV surveillance. Thus, within a portion of the border that is covered by a UAV, it is necessary to keep the UAV moving.

How to describe UAV patrolling strategies. For simplicity, let us assume that the UAV can fly reasonably fast along the border, so that for each point, the interval between two consequent overflies does not exceed the time $2T$ needed to successfully cross the border area back-and-forth.

In the ideal case, this would means that the UAV is capable of detecting all adversaries—and thus, preventing all border violations. In reality, however, a fast flying UAV can miss the adversary. It is therefore desirable to select a trajectory that would minimize the effect of this miss.

The faster the UAV goes pass a certain location, the less time it spends in the vicinity of this location, the more probable it is that the UAV will miss the adversary. From this viewpoint, an important characteristic of the trajectory is the velocity $v(x)$ with which the UAV passes through the location x. So, by a patrolling strategy, we will mean a function $v(x)$ that describes how fast the UAV flies at different locations.

This strategy must be selected in such a way that a total time for a UAV to go from one end of the area to another one is equal to the given value T. The time during which a UAV passes from the location x to the location $x + \Delta x$ is equal to

$$\Delta t = \frac{\Delta x}{v(x)}. \tag{2.3.1}$$

Thus, the overall flight time is equal to the sum of these times, i.e., to

$$T = \int \frac{dx}{v(x)}, \tag{2.3.2}$$

where the integral is taken over the whole length of the border segment.

From the mathematical viewpoint, an arbitrary non-negative function $v(x)$ can describe the velocity at different locations. In practice, not every function $v(x)$ can be implemented, since the UAV has the largest possible velocity V, so we must have $v(x) \le V$ for all x.

From the computational viewpoint, it is convenient, instead of the velocity $v(x)$, to use its reciprocal

$$s(x) \stackrel{\text{def}}{=} \frac{1}{v(x)}. \tag{2.3.3}$$

In the geosciences, this reciprocal is called *slowness*; see, e.g., [13] and references therein; we will use this term in this section as well.

In terms of slowness, the requirement that the overall time be equal to T has a simpler form

$$T = \int s(x)\,dx. \tag{2.3.4}$$

In terms of slowness $s(x)$, the velocity limitation

$$v(x) \le V \tag{2.3.5}$$

takes the form $s(x) \ge S$, where $S \stackrel{\text{def}}{=} \frac{1}{V}$. Since $s(x) \ge S$, the value $s(x)$ can be represented as $S + \Delta s(x)$, where $\Delta s(x) \stackrel{\text{def}}{=} s(x) - S$ satisfy the simpler constraint $\Delta s(x) \ge 0$.

In terms of $\Delta s(x)$, the requirement that the overall time be equal to T has a simpler form

$$T = S \cdot L + \int \Delta s(x)\,dx, \tag{2.3.6}$$

where L is the total length of the piece of the border that we are defending, or, equivalently,

$$T_0 = \int \Delta s(x)\,dx, \tag{2.3.7}$$

where $T_0 \stackrel{\text{def}}{=} T - S \cdot L$.

Probability of detection. In order to select a reasonable patrolling strategy, we must find out, for each strategy, what is the probability that under this strategy, the adversary can still cross the border.

Let h denote a distance at which the UAV can still see. This means that when the adversary is trying to cross at location x, a UAV can, in principle, observe this adversary when it is located in the zone between $x - h$ and $x + h$. The width of this zone is equal to

$$(x + h) - (x - h) = 2h. \tag{2.3.8}$$

We have denoted the UAV's velocity at location x by $v(x)$. So, the time that it takes for a UAV to cross the zone of width $2h$ is equal to

$$t_{\text{obs}} = \frac{2h}{v(x)}. \tag{2.3.9}$$

In terms of slowness, this expression takes a simpler form

$$t_{\text{obs}} = 2h \cdot s(x). \tag{2.3.10}$$

Let Δt denote the time during which a UAV takes one snapshot of the underlying area. In these terms, during the crossing time t_{obs}, the UAV can take

$$n(x) = \frac{t_{\text{obs}}}{\Delta t} = \frac{2h}{\Delta t} \cdot s(x) \tag{2.3.11}$$

snapshots.

Let p_1 be the probability that an adversary can avoid detection based on a single snapshot. Then, to avoid detection during several snapshots means to avoid detection during the first snapshot, during the second snapshot, etc. It is reasonable to assume that the misses corresponding to different snapshots are statistically independent. Under this assumption, the probability $p(x)$ to be missed under $n(x)$ snapshots is equal to the product of $n(x)$ probabilities of a miss corresponding to different snapshots, i.e., equal to

$$p(x) = p_1^{n(x)}. \tag{2.3.12}$$

Substituting the above expression for $n(x)$ in terms of $s(x)$, we conclude that

$$p(x) = p_1^{(2h/\Delta t)\cdot s(x)}, \tag{2.3.13}$$

i.e., that

$$p(x) = \exp(-k \cdot s(x)), \tag{2.3.14}$$

where we denoted

$$k \stackrel{\text{def}}{=} \frac{2h}{\Delta t} \cdot |\ln(p_1)|. \tag{2.3.15}$$

Relative importance of different locations. We also need to take into account that different locations along the border have different importance.

For example, if smugglers succeed in bringing drugs to the vicinity of the city of El Paso, they can store in a safe place and distribute it without exposure. On the other hand, if they bring the same shipment in the remote desert area, they still need to bring it close to a town or a city, and risk being detected while they are transporting this shipment.

In the case of smugglers, this importance can be described in monetary terms: a shipment available in city can be sold for a much larger amount than a shipment available at some remote location from which it still has to be transported to a city. The corresponding price $w(x)$ of the shipment successfully transported across the border at a point with coordinate x can be used as a measure of potential benefit, for the adversary, of penetrating the border at this particular location.

For other types of border penetration, we can also similarly estimate the potential benefit to the adversary.

We will start our analysis with a simplified case when we know the exact value of $w(x)$ for all x. After that, we will explain how to deal with a more realistic case, when we only know $w(x)$ with uncertainty.

Decision making: reminder. We assume that the adversary has observed the UAV, so the adversary knows the slowness function $s(x)$ and is, thus, capable of computing the probability $p(x)$ of avoiding detection. How does an adversary make decisions based on this knowledge?

A standard way to describe preferences of a decision maker is to use the notion of *utility*; see, e.g., [14–18]. To describe the utility of an outcome A, we need to select two extreme outcomes: a very unfavorable alternative A_- and a very favorable outcome A_+.

We assume that all outcomes A in which we are interested are better than A_- and worse than A_+. If we denote the relation "the decision maker prefers A' to A" by $A \le A'$, then we can describe this assumption as $A_- \le A \le A_+$.

Then, for each probability $p \in [0, 1]$, we can consider a *lottery* $L(p)$ in which we have A_+ with probability p and A_- with the remaining probability $1 - p$.

For $p = 1$, the lottery $L(p)$ coincides with A_+, so we have $A \le A(1)$. For $p = 0$, the lottery $L(p)$ coincides with A_-, so we have $A(0) \le A$. The larger p, i.e., the larger the probability of a beneficial event A_+, the more beneficial is the lottery $L(p)$ for the decision maker. So, if $p < q$, then $L(p) < L(q)$.

Let p_0 be the infimum (greatest lower bound) of the set of all the values p for which $A \le L(p)$. Then:

- When $p < p_0$, then for $\widetilde{p} = (p + p_0)/2$, we have $\widetilde{p} < p_0$ and thus, by definition of the infimum, we cannot have $A \le L(\widetilde{p})$. Thus, we have $L(\widetilde{p}) \le A$. Since $p < \widetilde{p}$, we have $L(p) < L(\widetilde{p}) \le A$ and thus, $L(p) < A$.
- When $p > p_0$, then, since p_0 is the greatest lower bound, p is not a lower bound, i.e., there exists a value $\widetilde{p}$ for which $A \le L(\widetilde{p})$ and $\widetilde{p} < p$. Since $\widetilde{p} < p$, we have $L(\widetilde{p}) < L(p)$ hence $A < L(p)$.

Thus, we have the value p_0 that has the following property:

- when $p < p_0$, the corresponding lottery is worse than the event A:

$$L(p) < A; \tag{2.3.16}$$

- when $p > p_0$, the corresponding lottery is better than the event A:

$$L(p) > A. \quad (2.3.17)$$

This threshold value p_0 is called the *utility* of the event A. The utility is usually denoted by $u(A)$.

We can simplify the above somewhat complicated relation between A and p_0 by saying that the event $L(p_0)$ is *equivalent* to A. We will denote this equivalence by $A \sim L(p_0)$.

The notion of utility depends on the choice of the outcomes A_- (for which utility is 0) and A_+ (for which utility is 1). In principle, we select different outcomes A'_- and A'_+. One can show that the new value $u'(A)$ is linearly related to the old one: $u'(A) = a \cdot u(A) + b$, where:

- $b = u'(A_-)$ is the utility of A_- in the new scale, and
- $a + b = u'(A_+)$ is the utility of A_+ in the new scale, so we can determine a as $u'(A_+) - u'(A_-)$.

In other words, utility is defined modulo an arbitrary linear transformation

$$u(A) \to u'(A) = a \cdot u(A) + b. \quad (2.3.18)$$

In practice, we can rarely predict the exact consequences of each decision. The consequences depend on the circumstances. For example, if we decide whether to take an umbrella or not, the consequences of this decision depend on whether it will rain or not. In the ideal situation, we know the probabilities $p_1, \ldots, p_n$ of different possible consequences $E_1, \ldots, E_n$. In other words, the action leads to E_1 with probability p_1, to E_2 with probability p_2, ..., and to E_n with probability p_n.

By definition of the utility, the event E_1 is equivalent to a lottery $L(u(E_1))$ in which we get A_+ with probability $u(E_1)$, the event E_2 is equivalent to a lottery $L(u(E_2))$ in which we get A_+ with probability $u(E_2)$, etc. Thus, the original action is equivalent to the composite lottery, in which:

- with probability p_1, we get a lottery that results in A_+ with probability $u(E_1)$, and in A_- otherwise;
- with probability p_2, we get a lottery that results in A_+ with probability $u(E_2)$, and in A_- otherwise;
- ...

In this composite lottery, we get either A_+ or A_-, and the probability of getting A_+ can be easily computed as

$$u \stackrel{\text{def}}{=} p_1 \cdot u(E_1) + p_2 \cdot u(E_2) + \cdots + p_n \cdot u(E_n). \quad (2.3.19)$$

Thus, the original action is equivalent to the lottery $L(u)$. By definition of the utility, this means that the utility of the action is equal to u.

From the mathematical viewpoint, u is the expected value of the utility of different consequences, so we can conclude that the utility of an action is the expected value of utilities of its consequences.

Strategy selected by the adversary. We have already mentioned that utility is defined modulo an arbitrary linear transformation. For convenience, let us select the utility scale in such a way that for the adversary, the utility of not being able to cross the border is 0.

In this scale, let $w(x)$ denote the utility of the adversary succeeding in crossing the border at location x. We have assumed that we know the exact value of $w(x)$ for every location x.

According to decision theory, the adversary will select a location x at which the expected utility

$$u(x) = p(x) \cdot w(x) = \exp(-k \cdot s(x)) \cdot w(x) \tag{2.3.20}$$

is the largest possible.

Thus, for each slowness function $s(x)$, the adversary's gain $G(s)$ is equal to

$$G(s) = \max_x u(x) = \max_x \left[\exp(-k \cdot s(x)) \cdot w(x)\right]. \tag{2.3.21}$$

Towards an optimal strategy for patrolling the border. Our objective is to select a strategy $s(x)$ for which the gain $G(s)$ is the smallest possible.

Let x_m be the location at which the utility $u(x) = \exp(-k \cdot s(x)) \cdot w(x)$ attains its largest possible value. If close to x_m, we have a point x_0 for which $u(x_0) < u(x_m)$ and $s(x_0) > S$, then we can slightly decrease the slowness $s(x_0)$ at the vicinity of x_0 (i.e., go faster in this vicinity) and use the resulting time to slow down (i.e., to go slower) at all locations x at which $u(x) = u(x_m)$. As a result, we slightly decrease the value $u(x_m) = \exp(-k \cdot s(x_m)) \cdot w(x_m)$.

Yes, we also slightly increase the value

$$u(x_0) = \exp(-k \cdot s(x_0)) \cdot w(x_0), \tag{2.3.22}$$

but for small changes, this value is still smaller that $u(x_m)$ and thus, does not affect the maximum $\max_x u(x)$. As a result, the gain $G(s)$ decreases (this argument is similar to the one presented in [19]).

So, when the adversary's gain is minimized, we get

$$u(x) = u_0 = \text{const} \tag{2.3.23}$$

hence

$$\exp(-k \cdot s(x)) = \frac{u_0}{w(x)}, \tag{2.3.24}$$

thence

$$s(x) = \frac{1}{k} \cdot (\ln(w(x)) - \ln(u_0)) \tag{2.3.25}$$

and

$$\Delta s(x) = \frac{1}{k} \cdot \ln(w(x)) - \Delta_0, \tag{2.3.26}$$

where

$$\Delta_0 \stackrel{\text{def}}{=} -\frac{1}{k} \cdot \ln(u_0) - S. \tag{2.3.27}$$

When this value gets to $s(x) = S$ and $\Delta s(x) = 0$, we get $\Delta s(x) = S$. Thus, we conclude that

$$\Delta s(x) = \max\left(\frac{1}{k} \cdot \ln(w(x)) - \Delta_0, 0\right). \tag{2.3.28}$$

The value Δ_0 can be determined from the condition that

$$\int \Delta s(x)\,dx =$$

$$\int \max\left(\frac{1}{k} \cdot \ln(w(x)) - \Delta_0, 0\right) dx = T_0. \tag{2.3.29}$$

Since this integral monotonically decreases with Δ_0, we can use bisection to find the appropriate value Δ_0; see, e.g., [20].

Towards taking fuzzy uncertainty into account . The above algorithm is based on the assumption that we know the exact value of the adversary's gain $w(x)$ at different locations. In reality, as we have mentioned, we only have expert estimates for $w(x)$. To formalize these estimates, we can use fuzzy techniques; see, e.g., [21, 22].

Once we have the fuzzy values $w(x)$, we can apply Zadeh's extension principle to the above crisp formulas and thus, come up with fuzzy recommendations about the slowness, such as "go somewhat slow here", "go fast", etc. It is well known (see, e.g., [21, 22]) that Zadeh's extension principle is equivalent to processing α-cuts. Specifically, if we know a relation $y = f(x_1, \ldots, x_n)$ between the inputs $x_1, \ldots, x_n$ and the desired value y, and we know the fuzzy values $X_1, \ldots, X_n$ of the inputs, then the resulting fuzzy value Y of the output can be obtained as follows: for every $\alpha \in (0, 1]$, we have

$$Y(\alpha) = f(X_1(\alpha), \ldots, X_n(\alpha)) =$$

$$\{f(x_1, \ldots, x_n) : x_1 \in X_1(\alpha), \ldots, x_n \in X_n(\alpha)\}, \tag{2.3.30}$$

where for each fuzzy value Z with a membership function $\mu_Z(z)$, its α-cut $Z(\alpha)$ is defined as

$$Z(\alpha) \stackrel{\text{def}}{=} \{z : \mu_Z(z) \geq \alpha\}. \tag{2.3.31}$$

When a fuzzy value is a fuzzy number, each α-cut is an interval $Z(\alpha) = [\underline{Z}(\alpha), \overline{Z}(\alpha)]$. When all the inputs are fuzzy numbers, the above formula takes the simplified form

$$[\underline{Y}(\alpha), \overline{Y}(\alpha)] = \{f(x_1, \ldots, x_n) : x_i \in [\underline{X}_i(\alpha), \overline{X}_i(\alpha)]. \tag{2.3.32}$$

When the function $y = f(x_1, \ldots, x_n)$ is an increasing function of all its variables, then its largest value is attained when all its inputs attain their largest values, and its smallest value is attained when all its inputs attain their smallest values. In other words, the desired α-cut has the form $[\underline{Y}(\alpha), \overline{Y}(\alpha)]$, where

$$\underline{Y}(\alpha) = f(\underline{X}_1(\alpha), \ldots, \underline{X}_n(\alpha)); \tag{2.3.33}$$

$$\overline{Y}(\alpha) = f(\overline{X}_1(\alpha), \ldots, \overline{X}_n(\alpha)). \tag{2.3.34}$$

Similarly, when the function $y = f(x_1, \ldots, x_n)$ is an increasing function of the variables $x_1, \ldots, x_k$ and decreasing in $x_{k+1}, \ldots, x_n$), then the α-cut has the form $[\underline{Y}(\alpha), \overline{Y}(\alpha)]$, where

$$\underline{Y}(\alpha) = f(\underline{X}_1(\alpha), \ldots, \underline{X}_k(\alpha), \overline{X}_{k+1}(\alpha), \ldots, \overline{X}_n(\alpha)); \tag{2.3.35}$$

$$\overline{Y}(\alpha) = f(\overline{X}_1(\alpha), \ldots, \overline{X}_k(\alpha), \underline{X}_{k+1}(\alpha), \ldots, \overline{X}_n(\alpha)). \tag{2.3.36}$$

In our case, for each location x, we know the fuzzy value $W(x)$ of the corresponding gain. This means that for each degree α, we know the corresponding α-cut $W(x)(\alpha) = [\underline{W(x)}(\alpha), \overline{W(x)}(\alpha)]$.

In the crisp case, based on the gains $w(x)$, we first compute the value Δ_0 and then the corresponding changes $\Delta s(x)$ in the UAV's slowness. Thus, in the fuzzy case, we need to find the α-cuts for Δ_0 and then, α-cuts for $\Delta s(x)$.

According to the above formula for Δ_0, its value is an increasing function of all the inputs $w(x)$. Thus, we conclude that for every α, the α-cut for Δ_0 has the form $[\underline{\Delta}_0(\alpha), \overline{\Delta}_0(\alpha)]$, where $\underline{\Delta}_0(\alpha)$ can be determined from the condition that

$$\int \max\left(\frac{1}{k} \cdot \ln(\underline{W(x)}(\alpha)(x)) - \underline{\Delta}_0(\alpha), 0\right) dx = T_0, \tag{2.3.37}$$

and $\overline{\Delta}_0(\alpha)$ can be determined from the condition that

$$\int \max\left(\frac{1}{k} \cdot \ln(\overline{W(x)}(\alpha)(x)) - \overline{\Delta}_0(\alpha), 0\right) dx = T_0. \tag{2.3.38}$$

The value $\Delta s(x)$ is increasing in $w(x)$ and decreasing in Δ_0. Thus,

- the smallest value $\underline{\Delta s(x)}(\alpha)$ is attained when $w(x)$ is the smallest and Δ_0 is the largest, and

- the largest value $\overline{\Delta s(x)}(\alpha)$ is attained when $w(x)$ is the largest and Δ_0 is the smallest:

$$\underline{\Delta s(x)}(\alpha) = \max\left(\frac{1}{k} \cdot \ln(\underline{W(x)}(\alpha)) - \overline{\Delta}_0(\alpha), 0\right); \quad (2.3.39)$$

$$\overline{\Delta s(x)}(\alpha) = \max\left(\frac{1}{k} \cdot \ln(\overline{W(x)}(\alpha)) - \underline{\Delta}_0(\alpha), 0\right). \quad (2.3.40)$$

The resulting recommendations can be used either as a guidance for a human controller, or—by using fuzzy control—in the design of the automatic controller.

Comment. Fuzzy techniques can be similarly used in other problems related to security, e.g., in finding optimal placement for bio-weapon detectors [5] as described in Sect. 2.1.

2.4 Efficient Algorithms for Optimizing Sensor Use: Case Study of Security Problems

In this section, we analyze the problem of designing efficient algorithms for optimizing sensor use, on the example of security problems.

The results from this section first appeared in [23].

Formulation of the problem. Security problems typically involve making strategic resource allocation decisions in order to prevent or mitigate attacks. Game theory has been used to model decision-making in a variety of security situations, including the protection of critical infrastructure from terrorist attacks [24, 25], computer network security [26–28], robot patrolling [29–31], and scheduling [32]. Recently, research on security games has been deployed to make real-world homeland security decision, including the ARMOR system in use at the LAX airport [33], the IRIS system used by the Federal Air Marshals Service [34], and the GUARDS system developed for the Transportation Security Administration [35].

A key research direction has been the development of faster algorithms to scale to increasingly large and complex instances of security games [19, 36–38]. Faster algorithms that exploit the structure of security games have been key in enabling new applications of these methods. We present new algorithms for one of the most basic classes of security games: Stackelberg security games with multiple, identical defender resources. This class of games was described by Kiekintveld et al. [19], which also gave a polynomial-time ($O(n^2)$) algorithm for computing Stackelberg equilibrium of these games.

In this section, we present two new algorithms for Stackelberg security games with identical resources. The first solves a special case in worst-case linear time ($O(n)$), and the second solves the general case in $O(n \cdot \log(n))$. In addition to improving on the theoretical complexity of the best known methods for this class of security games, our algorithms are based on a detailed analysis of the structure of the solutions for these games, which may lead to faster algorithms or heuristics for more complex variants of security games.

Security game model: general case. In this section, we adopt the general model of security games described in [19]. A security game has two players, a *defender*, Θ, and an *attacker*, Ψ. There is a set of n targets $t_i \in T$ that the attacker wishes to attack and the defender wishes to protect. In our model, the attacker can choose to attack exactly one target from this set. The defender has a limited number of resources, $m < n$, that can be deployed to protect the targets. We assume throughout that these resources are identical, and that at most one resource can be used to protect each target.

If the attacker chooses to attack target t_i, we call the attack *successful* if the target is left uncovered by a defender, and *unsuccessful* if the target is covered by a defender. The defender's payoff for an uncovered attack is denoted $U^u_\Theta(t)$, and for a covered attack $U^c_\Theta(t)$. Similarly, $U^u_\Psi(t)$ and $U^c_\Psi(t)$ denote the attacker's payoffs in each case. We will make the standard assumptions for security games that $U^u_\Theta(t) < U^c_\Theta(t)$ and $U^u_\Psi(t) > U^c_\Psi(t)$ for all targets t. In other words, the attacker receives a higher payoff for attacking an undefended target than a defended one, and vice versa for the defender. Note that this does not imply that the games are zero-sum (or even strategically zero-sum).

The attacker's possible strategies consist of attacking each of the n targets. The defender's space of possible strategies consists of all possible ways to assign the m resources to the n targets. However, we can conveniently summarize the defenders strategy by defining the *coverage vector* which gives the probability that there is a defender resource assigned to each individual target. Denote these probabilities by c_i, so that $\sum_{i=1}^{n} c_i = m$. The attacker's expected utility for an attack on target t_i can then be written as $(1 - c_i) \cdot U^u_\Psi(t_i) + c_i \cdot U^c_\Psi(t_i)$, and similarly for the defender. Because of our assumptions, for each target t_i, the defender's expected payoff decreases when the probability c_i of defending this target increases. We also assume that all defender resources are identical and can be deployed to any target.

We model the game as a *Stackelberg game* [39] in which the attacker can observe the defender's strategy $(c_1, \ldots, c_n)$ before planning an attack (modeling the capability of attackers to use surveillance to learn security policies). The standard solution concept for these games is Strong Stackelberg Equilibrium (SSE) [40, 41]. In an SSE, the leader first selects a mixed strategy, and then the follower chooses an optimal pure strategy in response, breaking ties in favor of the leader. This behavior can be induced by the leader selecting a strategy arbitrarily close to the equilibrium that causes the the follower to strictly prefer the desired strategy [42], but in practice we compute the limit point where ties are broken in favor of the leader.

Case of fully protective resources: description and analysis of the problem. Let us first consider a practically important case of fully protective resources. When a single resource is deployed at a target, the target is fully protected. For now, we restrict the attacker's payoff for attacking a covered resource to 0: $U^c_\Psi(t) = 0$. We begin with a basic analysis that describes the structure of the solution.

According to the analysis in [19], in our game, the objective of the defender is equivalent to minimizing the expected utility of the attacker, as long as the solution has the largest possible set of targets that are optimal for the attacker to select. Using the Strong Stackelberg Equilibrium assumption, the attacker will select breaking ties in favor of the defender. Therefore, in most cases we will not need to take into account the defender's payoffs directly; the defender's payoff will be maximized implicitly by finding a set of coverage probabilities c_i so as to minimize the expected payoff of the attacker.

The attacker seeks to maximize the expected value of a successful attack:

$$\arg\max_i \, (1 - c_i) \cdot U^u_\Psi(t_i), \tag{2.4.1}$$

while the defender chooses a coverage vector to minimize the attacker's expected payoff.

Let t_{i_o} denote the optimal target to attack, so we have for every target t_i:

$$(1 - c_{i_o}) \cdot U^u_\Psi(t_{i_o}) \geq (1 - c_i) \cdot U^u_\Psi(t_i). \tag{2.4.2}$$

Now, assume that for some i this inequality is strict and that $c_i > 0$. In this case we could decrease c_i and increase the probability c_j for all j such that

$$(1 - c_j) \cdot U^u_\Psi(t_j) = (1 - c_{i_o}) \cdot U^u_\Psi(t_{i_o}), \tag{2.4.3}$$

thus decreasing the expected payoff of the attacker.

Therefore, for the minimizing coverage vector, all targets can be divided into two groups:

- either the expected value for attacking the target is equal to the optimal value,
- or the expected value is less than the optimal value and the coverage probability assigned to the target is 0.

In other words, the optimal solution will have the property that the attacker's expected value for all targets with positive coverage probability is equal to some constant q:

$$(1 - c_i) \cdot U^u_\Psi(t_i) = q. \tag{2.4.4}$$

For any target t_i with $c_i > 0$ we can thus calculate the necessary value of c_i as:

$$c_i = 1 - \frac{q}{U^u_\Psi(t_i)}. \tag{2.4.5}$$

For all other targets $U^u_\Psi(t_i) < q$, and therefore

$$1 - \frac{q}{U^u_\Psi(t_i)} < 0. \tag{2.4.6}$$

Summarizing: once we know q, we can find all the probabilities c_i by using the formula

$$c_i = \max\left(1 - \frac{q}{U^u_\Psi(t_i)}, 0\right). \tag{2.4.7}$$

For each target t_i, this formula requires a constant number of computational steps. Therefore, after q is computed, we can therefore compute all the probabilities c_i in time $O(n)$.

So, to find the optimal covering vector, it is sufficient to find the constant q. This constant can be found from the condition that $\sum\limits_{i=1}^{n} c_i = m$, i.e., that

$$\sum_{i=1}^{n} \max\left(1 - \frac{q}{U^u_\Psi(t_i)}, 0\right) = m. \tag{2.4.8}$$

The left-hand side of this equality decreases as q increases. So, if for some q, the resulting sum is smaller than m, this means that the optimal value q_o is smaller than q: $q_o < q$; similarly, if for some q, the resulting sum is larger than m, this means that the optimal value q_o is larger than q: $q_o > q$.

The structure of the optimal covering vector can be clarified if we sort the targets in order of descending attacker payoffs for successful attacks, so that:

$$U^u_\Psi(t_{(1)}) \geq \cdots \geq U^u_\Psi(t_{(n-1)}) \geq U^u_\Psi(t_{(n)}). \tag{2.4.9}$$

We can also add $U^u_\Psi(t_{(0)}) \stackrel{\text{def}}{=} +\infty$ and $U^u_\Psi(t_{(n+1)}) \stackrel{\text{def}}{=} 0$, then

$$U^u_\Psi(t_{(0)}) \geq \cdots \geq U^u_\Psi(t_{(n)}) \geq U^u_\Psi(t_{(n+1)}). \tag{2.4.10}$$

The values $U^u_\Psi(t_{(i)})$ divide the real line into intervals $[U^u_\Psi(t_{(i+1)}), U^u_\Psi(t_{(i)})]$, so the threshold constant q must be in one of these intervals, i.e., between $U^u_\Psi(t_{(k+1)})$ and $U^u_\Psi(t_{(k)})$ for some k. In this case, according to the above formula for c_i, all targets with a value greater than q (i.e., the targets $t_{(1)}$, $t_{(2)}$, ..., $t_{(k)}$ in the above ordering) will be protected with positive probability, and all targets with value smaller than q (i.e., targets $t_{(k+1)}, t_{(k+2)}, \ldots$) are left unprotected. Let k denote the index of last target that has positive probability. Given the constraint that the coverage probabilities add to m, we can write:

$$\sum_{i=1}^{k}\left(1 - \frac{q}{U^u_\Psi(t_{(i)})}\right) = m, \tag{2.4.11}$$

hence

$$k - m = q \cdot \sum_{i=1}^{k} \frac{1}{U_\Psi^u(t_{(i)})}, \tag{2.4.12}$$

and

$$q = \frac{k - m}{\sum_{i=1}^{k} \frac{1}{U_\Psi^u(t_{(i)})}}. \tag{2.4.13}$$

So, instead of selecting q, we can simply select a threshold value k.

Once we have found this k, we can then compute the threshold value q by using the above formula and then use this q to find the optimal coverage probabilities.

For the optimal value $k = k_o$, the corresponding value q is located in the interval $[U_\Psi^u(t_{(k+1)}), U_\Psi^u(t_{(k)})]$. If for some k, the value q computed by the above formula is smaller than $U_\Psi^u(t_{(k+1)})$, this means that we are trying to cover too few targets, with the same q, we can cover more, so the optimal value k_o should be larger: $k > k_o$.

Similarly, if for some k, the value q computed by the above formula is larger than $U_\Psi^u(t_{(k)})$, this means that we are trying to cover too many targets, so the optimal value k_o should be smaller: $k_o < k$.

Let us show that this argument can lead to a linear-time algorithm for finding the optimal coverage vector.

Case of fully protective resources: linear-time algorithm. On each stage of this iterative algorithm, we have three lists of targets:

- the list T^c of the targets t_i about which we are sure that in the optimal coverage, these targets will be covered with a positive probability $c_i > 0$;
- the list T^u of the targets t_i about which we are sure that in the optimal coverage, these targets will not be covered ($c_i = 0$);
- the list $T^?$ of the targets t_i about which we have not yet found out whether they will be covered or not in the optimal coverage.

In the beginning, we set $T^c = T^u = \emptyset$ and

$$T^? = \{t_1, t_2, \ldots, t_n\}. \tag{2.4.14}$$

At each stage, we will also update the value

$$S^c = \sum_{t_i \in T^c} \frac{1}{U_\Psi^u(t_i)}. \tag{2.4.15}$$

In the beginning, since $T^c = \emptyset$, we take $S^c = 0$.

At each iteration, we do the following:

- First, we compute the median m of the values $U_\Psi^u(t_i)$ corresponding to all "undecided" targets $t_i \in T^?$.

- Then, by analyzing the elements of the undecided set $T^?$ one by one, we divide them into two subsets

$$T^+ = \{t_i : U^u_\Psi(t_i) \geq m\}, T^- = \{t_i : U^u_\Psi(t_i) < m\}. \tag{2.4.16}$$

In the set T^+, we find the target t_{k^+} with the smallest value of $U^u_\Psi(t_i)$; in the set T^-, we find the target t_{k^-} with the largest value of $U^u_\Psi(t_i)$.
- We then compute

$$S^+ = \sum_{t_i \in T^+} \frac{1}{U^u_\Psi(t_i)}, \tag{2.4.17}$$

$s^+ = S^c + S^+$, and $q = \dfrac{k-m}{s^+}$.
- If $q < U^u_\Psi(t_{k^-})$, then, as we have argued in our analysis, this means that we are trying to cover too few targets, so definitely all the elements from the set T^+ should be covered. Thus, we replace T^c with $T^c \cup T^+$, $T^?$ with T^-, and S^c with s^+.
- If $q > U^u_\Psi(t_{k^+})$, this means that we are trying to cover too many targets, so definitely all the elements from the set T^- should not be covered. Thus, we replace T^u with $T^u \cup T^-$ and $T^?$ with T^+ (and keep S^c unchanged).
- Finally, if $U^u_\Psi(t_{k^-}) \leq q \leq U^u_\Psi(t_{k^+})$, this means that this q is optimal.

Iterations continue until we find the optimal value q. Once we get the optimal value q, we can then find the optimal covering probabilities as $c_i = \max\left(1 - \dfrac{q}{U^u_\Psi(t_i)}, 0\right)$.

Let us prove that this algorithm indeed takes linear time. Indeed, at each iteration, we can compute the median in linear time [20], and all other operations with the set $T^?$ also take time $\mathscr{T}$ linear in the number of elements $|T^?|$ of this set $T^?$: $\mathscr{T} \leq C \cdot |T^?|$ for some C. We start with the set $T^?$ of size n. On the next iteration, we have a set of size $n/2$, then $n/4$, etc. Thus, the overall computation time is

$$\leq C \cdot (n + n/2 + n/4 + \cdots) \leq C \cdot 2n,$$

i.e., linear in n.

General case: analysis of the problem. Let us now go back to the general case, when defense resources are not necessarily fully protective. In this general case, the attacker seeks to maximize the expected value of a successful attack:

$$\arg\max_\ell e_\ell(c_\ell), \tag{2.4.18}$$

where

$$e_\ell(c_\ell) \stackrel{\text{def}}{=} (1 - c_\ell) \cdot U^u_\Psi(t_\ell) + c_\ell \cdot U^c_\Psi(t_\ell), \tag{2.4.19}$$

while the defender chooses a coverage vector to minimize the attacker's expected payoff

$$e(c) = \max_{\ell} e_\ell(c_\ell). \tag{2.4.20}$$

Once we select a coverage vector, we thus divide all the targets into three groups:

- the first group is formed by targets t_i for which $c_i = 1$; these targets that will be guarded with certainty;
- the second group is formed by targets t_j for which $0 < c_j < 1$; these targets with some probability will be guarded and with some probability will not be guarded;
- the third group is formed by targets t_k for which $c_k = 0$; these targets will be not guarded.

Intuitively, this division makes sense:

- the most important targets must be guarded no matter what,
- the least valuable targets will not be guarded at all if we do not have enough resources, and
- intermediate targets will be guarded with some probability.

Let us prove that this intuitive meaning is indeed true. To be more precise, let us prove that in this game, there exists a minimizing vector $(c_1, \ldots, c_n)$ that has the following properties:

- The expected payoff $e_(c_i)$ of each target t_i of the first group (with $c_i = 1$) is larger than or equal to the expected payoff $e_j(c_j)$ of each target t_j of the second group (with $0 < c_j < 1$):

$$e_i(c_i) \ge e_j(c_j). \tag{2.4.21}$$

- The expected payoff $e_j(c_j)$ of all target $t_j, t_{j'}$ from the second group (with $0 < c_j < 1$) is the same:

$$e_j(c_j) \ge e_{j'}(c_{j'}). \tag{2.4.22}$$

- The expected payoff $e_j(c_j)$ of each target t_j from the second group (with $0 < c_j < 1$) is larger than or equal to the expected payoff of each target t_k from the third group (with $c_k = 0$):

$$e_j(c_j) \ge e_k(c_k). \tag{2.4.23}$$

Intuitively, this makes sense: if the attacker's expected payoff from a target t_i that we guard absolutely is smaller than the expected payoff from some other target t_j that we guard with a certain probability, then it makes sense to switch some probability from target t_i to target t_j. In this case, the attacker's expected value for t_j decreases; for t_i it somewhat increases, but since it was smaller than for the target t_j, it remains smaller, and the maximum of these values $e_i(c_i)$ does not increase.

To prove this result more formally, let us start with any minimizing vector and show that by appropriate transformations it can be transformed into a minimizing vector with the desired properties.

First, let us show how we can satisfy the first property. For that, let us show that we can decrease the number of targets t_i for which $c_i = 1$ and for which, for some j, we have $0 < c_j < 1$ and $e_i(c_i) < e_j(c_j)$. Indeed, out of all such targets, let us pick a target for which the value $e_i(c_i)$ is the smallest, and let j be the corresponding target from the second group. Then, for some $\Delta > 0$, we replace c_i with $c'_i = c_i - \Delta$ and c_j with $c'_j = c_j + \Delta$. When Δ is small enough, we have $c'_i > 0$, $c'_j < 1$, and $e_i(c'_i)$ is still smaller than all the values $e_\ell(c_\ell)$ for which we had $e_i(c_i) < e_\ell(e_\ell)$.

Let us keep all the other probabilities the same: $e'_\ell = c_\ell$ for all $\ell \neq i, j$. This replacement does not change the sum $\sum c_i$, so while $c'_i \geq 0$ and $c'_j \leq 1$, we still get a coverage vector. As we have mentioned, the expected value of a target decreases with the increase in the probability that this target will be guarded. Thus, when Δ increases, the value $e_i(c_i - \Delta)$ increases while the value $e_j(c_j + \Delta)$ decreases. So, while $e_i(c_i - \Delta) \leq e_j(c_j + \Delta)$, we have $e_i(c_i) < e_i(c_i - \Delta) \leq e_j(c_j + \Delta) < e_j(c_j)$. Thus, $e_i(c'_i) < e_j(c_j) \leq e(c) = \max\limits_\ell c_\ell(e_\ell)$ and similarly $e_j(c'_i) < e_j(c_j) \leq e(c) = \max\limits_\ell c_\ell(e_\ell)$. For all other targets ℓ, we have $c'_\ell = c_\ell$ hence $e_\ell(c'_i) = e_\ell(c_\ell) \leq e(c)$. Thus,

$$e(c') = \min\left(e_i(c'_i), e_j(c'_j), \min_{\ell \neq i,j} e_\ell(c'_\ell))\right) \leq e(c). \tag{2.4.24}$$

Since the original vector c is a minimizing vector, the value $e(c)$ is the smallest possible value, we conclude that c' is also a minimizing vector.

Let us show that in the new minimizing vector, the number of targets ℓ from the first group for which the expected value is smaller than for some target from the second group is smaller than the same number computed based on the original minimizing vector. Indeed, in the new minimizing vector, the target t_i is no longer from group one, it is now from group two, so it is sufficient to check that this addition of a new group-two target does not lead to the appearance of a new "wrong-order" target of group one. Indeed, if for some target $t_{i'}$ from group one, we have $e_{i'}(c_{i'}) < e_i(c'_i)$, then we could not have $e_i(c_i) < e_{i'}(c_{i'})$—because we selected Δ so small that all such inequalities remain. Thus, we have $e_{i'}(c_{i'}) \leq e_i(c_i)$ but in this case $e_i(c_i) < e_j(c_j)$ implies that $e_{i'}(c_{i'}) < e_j(c_j)$—and thus, $t_{i'}$ was the wrong-order target already in the original minimizing vector.

By applying this procedure again and again, we arrive at the new minimizing vector for which the number of wrong-order targets of group one is 0, i.e., in which the expected payoff for every target from group one is larger than or equal to the expected payoff for every target from group two.

Similarly, we can get a new minimizing vector in which the expected payoff for every target from group two is larger than or equal to the expected payoff of every target of group three.

Let us now show that we can arrive at the minimizing vector for which for all targets from group two, the expected payoff is the same. Let us show how an appropriate procedure can minimize the number of pairs $(t_j, t_{j'})$ of targets from group two for which $e_j(c_j) < e_{j'}(c_{j'})$. Indeed, let us sort all the corresponding values $e_j(c_j)$ into an increasing sequence, and let us take two neighboring values from this sequence.

Similarly to the above case, we replace c_j with $c'_j = c_j - \Delta$ and $c_{j'}$ with $c'_{j'} = c_{j'} + \Delta$. Both expected values $e_j(c_j - \Delta)$ and $e_{j'}(c_{j'} + \Delta)$ linearly depend on Δ, so, by solving the corresponding linear equation, we can find Δ for which $e_j(c_j - \Delta) = e_{j'}(c_{j'} + \Delta)$. If this value Δ satisfies the conditions $c'_j = c_j - \Delta \geq 0$ and $c'_{j'} = c_{j'} + \Delta \leq 1$, we get a new minimizing vector in which strict inequality holds for one fewer pair of targets from group two. If this value Δ does not satisfy one of these inequalities, this means that for some smaller value $\Delta' < \Delta$, we have either $c'_j = 0$ or $c'_{j'} = 1$. In both cases, the pairs stops being a wrong-order pair of targets from group two. One can check that no other wrong-order pairs appear after this transformation.

Let us now take the minimizing vector with the desired properties. In particular, this means that for all targets from group two, the attacker's expected value is the same. Let us denote this common value by q. Then, for every target t_j with $0 < c_j < 1$, we have

$$(1 - c_j) \cdot U^u_\Psi(t_j) + c_\ell \cdot U^c_\Psi(t_j) = q. \tag{2.4.25}$$

So, we can calculate c_j as

$$c_j = \frac{U^u_\Psi(t_j) - q}{U^u_\Psi(t_j) - U^c_\Psi(t_j)}. \tag{2.4.26}$$

For targets for which $U^u_\Psi(t_k) < q$, we have $c_k = 0$—and the above ratio is negative. For targets for which $U^c_\Psi(t_i) > q$, we have $c_i = 1$—and the above ratio is larger than 1. Thus, if the ratio is smaller than 0, we take $c_i = 0$, and if the ratio is larger than 1, we take $c_i = 1$.

So, once we know q, for all targets t_i, we can find all the covering probabilities c_i by using the following formula:

$$c_i = \min\left(\max\left(\frac{U^u_\Psi(t_i) - q}{U^u_\Psi(t_i) - U^c_\Psi(t_i)}, 0\right), 1\right). \tag{2.4.27}$$

For each target t_i, this formula requires a constant number of computational steps. Therefore, after q is computed, we can therefore compute all the probabilities c_i in time $O(n)$.

So, to find the optimal covering vector, it is sufficient to find the constant q. This constant can be found from the condition that $\sum\limits_{i=1}^{n} c_i = m$, i.e., that

$$\sum_{i=1}^{n} \min\left(\max\left(\frac{U^u_\Psi(t_i) - q}{U^u_\Psi(t_i) - U^c_\Psi(t_i)}, 0\right), 1\right) = m. \tag{2.4.28}$$

The left-hand side of this equality decreases as q increases. So:

- If for some q, the resulting sum is smaller than m, this means that the optimal value q_o is smaller than q: $q_o < q$.
- Similarly, if for some q, the resulting sum is larger than m, this means that the optimal value q_o is larger than q: $q_o > q$.

Here, the target t_i is covered with probability $c_i > 0$ if and only if $q < U^u_\Psi(t_i)$, and the target t_i is covered with probability $c_i = 1$ if and only if $U^c_\Psi(t_i) \geq q$. Thus, the above formula for determining q can be rewritten as follows:

$$k(q) + \sum_{i:U^c_\Psi(t_i)<q\leq U^u_\Psi(t_i)} \frac{U^u_\Psi(t_i) - q}{U^u_\Psi(t_i) - U^c_\Psi(t_i)} = m, \tag{2.4.29}$$

where

$$k(q) \stackrel{\text{def}}{=} \#\{i : U^c_\Psi(t_i) \geq q\}. \tag{2.4.30}$$

Thus, if we know the place of q with respect to all the values $U^u_\Psi(t_i)$ and $U^c_\Psi(t_i)$, we can determine q by explicitly solving the above linear equation.

If we sort all $2n$ values $U^u_\Psi(t_i)$ and $U^c_\Psi(t_i)$ into a decreasing sequence

$$z_0 = +\infty \geq z_1 \geq z_2 \geq \ldots \geq z_{2n-1} \geq z_{2n} \geq z_{2n+1} = 0, \tag{2.4.31}$$

we thus subdivide the real line into $2n + 1$ zones $[z_{k+1}, z_k]$, within each of which the relation between q and the values $U^u_\Psi(t_i)$ and $U^c_\Psi(t_i)$ is fixed. Thus, within each zone, we can find the corresponding q and check whether this value is indeed within the corresponding zone. As a result, in order to find q, it is sufficient to find the corresponding value k.

Since the order is decreasing, the larger k, the smaller q, and the more targets we cover. The selection of the zone means that we select which targets we cover fully, and which targets we cover with a positive probability. Similar to the case of fully protective resources:

- If based on this selection, we need more than m resources—i.e., if the value q obtained from solving the above linear equation is smaller than all the values from this zone—this means that we are trying to cover too many targets, so we need to decrease k.
- Similarly, if it turns out that based on this selection, we need fewer than m resources—i.e., that the value q obtained from solving the above linear equation is larger than all the values from this zone—this means that we are trying to cover too few targets, so we can increase k.

Thus, we can use bisection to find the appropriate zone, and we arrive at the following algorithm.

General case: $O(n \cdot \log(n))$ algorithm. First, we sort all $2n$ values $U^u_\Psi(t_i)$ and $U^c_\Psi(t_i)$ into a decreasing sequence:

$$z_1 \geq z_2 \geq \cdots \geq z_{2n-1} \geq z_{2n}. \tag{2.4.32}$$

We then take $z_0 = +\infty$ and $x_{2n+1} = 0$, so that we get:

$$z_0 \geq z_1 \geq z_2 \geq \cdots \geq z_{2n-1} \geq z_{2n} \geq z_{2n+1}. \tag{2.4.33}$$

Then, we use bisection to find the value k for which $z_k \geq q \geq z_{k+1}$. At each stage of this bisection procedure, we keep two values ℓ and u such that $z_\ell \geq q \geq z_u$. In the beginning, we have $\ell = 0$ and $u = 2n + 1$. At each iteration, we do the following:

- First, we compute the midpoint $m = (\ell + u)/2$.
- Then, under the assumption that $q \in [z_{m+1}, z_m]$, we compute

$$k_m = \#\{i : U^c_\Psi(t_i) \geq z_{m+1}\}, \tag{2.4.34}$$

then $m_0 = m - k_m$, and find q from the resulting linear equation

$$\sum_{i : U^c_\Psi(t_i) \leq z_m \leq z_{m+1} \leq U^u_\Psi(t_i)} \frac{U^u_\Psi(t_i) - q}{U^u_\Psi(t_i) - U^c_\Psi(t_i)} = m_0. \tag{2.4.35}$$

- If the resulting value q is smaller than z_m, then, according to our analysis, this means that the optimal k is larger than m, so we replace the original value ℓ with m.
- If the resulting value q is larger than z_{m+1}, then, according to our analysis, this means that the optimal k is smaller than m, so we replace the original value u with m.

The algorithm stops when $u = \ell + 1$, in which case we have the desired q. Based on this q, we can compute all coverage probabilities by using the above formula

$$c_i = \min\left(\max\left(\frac{U^u_\Psi(t_i) - q}{U^u_\Psi(t_i) - U^c_\Psi(t_i)}, 0\right), 1\right). \tag{2.4.36}$$

There is one more special case the must be considered to ensure that this solution is in fact a Strong Stackelberg Equilibrium. This case occurs when at least one target has coverage $c_i = 1$. In this case, we must ensure that the target that gives maximum payoff for the defender has an optimal payoff for the attacker (so far, we have considered only the payoffs for the attacker). This can be done by first finding the maximal covered payoff for the attacker $U^c_\Psi(t)$ for any target that has coverage probability 1. Denote this target by t_{max}. We then loop through each of the targets to determine whether the defender would achieve a higher payoff if the coverage probability was reduced so that the attackers expected payoff was equal to $U^c_\Psi(t_{max})$.

We can compute the necessary coverage for each target using the equation:

$$c_i = \frac{U_\Psi^c(t_{max}) - U_\Theta^u(t_i)}{U_\Theta^c(t_i) - U_\Theta^u(t_i)} \tag{2.4.37}$$

If the defender's expected payoff for target t_i is greater than $U_\Psi^c(t_{max})$ given c_i, then we reduce the coverage probability to this new value c_i for target t_i. Note that this can only reduce the total coverage probability required. The additional coverage can either be left unallocated or assigned arbitrarily to any target for which the attacker has an expected payoff less than $U_\Psi^c(t_{max})$.

Let us prove that this algorithm indeed takes time

$$O(n \cdot \log(n)). \tag{2.4.38}$$

Indeed, sorting can be done in time $O(n \cdot \log(n))$ [20]. At each stage of the bisection method, we handle each target once, so each stage takes $O(n)$ computational steps. We start with an interval $[\ell, u]$ of size $2n$. At each stage, we replace it with a half-size interval $[\ell, m]$ or $[m, u]$. Thus, after the first iteration, we get an interval of size n, after the second, of size $n/2$, ..., and after k-th iteration, an interval of size $(2n)/2^k$. Thus, this procedure stops after $\log_2(2n)$ iterations. So, the overall computation time is indeed

$$O(n \cdot \log(n)) + O(n) \cdot \log(2n) = O(n \cdot \log(n)). \tag{2.4.39}$$

The final stage or analysis for the special case where at least one target coverage $c_i = 1$ requires two loop through each target. The first identifies the fully-covered target with maximum payoff for the attacker $U_\Psi^c(t_{max})$. The second calculates the required reduction in coverage probability to make the attacker indifferent between t_{max} and any other target, and replaces the coverage probability if a reduction is beneficial for the defender. Since this requires time $O(2 \cdot n)$, the overall complexity remains $O(n \cdot \log(n))$.

Summary of the results. In this section, we have presented two new algorithms for a fundamental class of Stackelberg security games. These algorithms operate in linear time for a restricted case, and $O(n \cdot \log(n))$ for the general case, both improvements over the best known algorithms for this class of games. The algorithms are based on new analysis of the structure of the game-theoretic solutions of these games, which may provide insights to improve the efficiency of algorithms for additional classes of security games.

References

1. B. Baesens, *Analytics in a Big Data World: The Essential Guide to Data Science and its Applications* (Wiley, Hoboken, New Jersey, 2014)
2. B. Marr, *Big Data: Using SMART Big Data, Analytics and Metrics To Make Better Decisions and Improve Performance* (Wiley, Chichester, UK, 2015)
3. N. Marz, J. Warren, *Big Data: Principles and Best Practices of Scalable Realtime Data Systems* (Manning Publication, Shelter Island, New York, 2015)
4. V. Mayer-Schönberger, K. Cukier, *Big Data: A Revolution That Will Transform How We Live, Work, and Think* (John Murray Publication, London, UK, 2013)
5. C. Kiekintveld, O. Lerma, Towards optimal placement of bio-weapon detectors, in *Proceedings of the 30th Annual Conference of the North American Fuzzy Information Processing Society NAFIPS'2011*, El Paso, Texas, 18–20 Mar 2011
6. R. Kershner, The number of circles covering a set. Am. J. Math. **61**(3), 665–671 (1939)
7. R.B. Myerson, *Game Theory: Analysis of Conflict* (Harvard University Press, Harvard, Massachusesst, 1997)
8. M. Koshelev, O. Lerma, C. Tweedie, Towards optimal few-parametric representation of spatial variation: geometric approach and environmental applications. Geombinatorics **21**(1), 15–24 (2011)
9. D.J. Sheskin, *Handbook of Parametric and Nonparametric Statistical Procedures* (Chapman & Hall/CRC, Boca Raton, Florida, 2011)
10. S. Li, Y. Ogura, V. Kreinovich, *Limit Theorems and Applications of Set Valued and Fuzzy Valued Random Variables* (Kluwer Academic Publishers, Dordrecht, 2002)
11. H. Busemann, *Geometry of Geodesics* (Dover, New York, 2005)
12. J. Aczel, *Lectures on Functional Equations and Their Applications* (Dover, New York, 2006)
13. M.G. Averill, K.C. Miller, G.R. Keller, V. Kreinovich, R. Araiza, S.A. Starks, Using expert knowledge in solving the seismic inverse problem. Int. J. Approx. Reason. **45**(3), 564–587 (2007)
14. P.C. Fishburn, *Utility Theory for Decision Making* (Wiley, New York, 1969)
15. P.C. Fishburn, *Nonlinear Preference and Utility Theory* (The John Hopkins Press, Baltimore, Maryland, 1988)
16. R.L. Keeney, H. Raiffa, *Decisions with Multiple Objectives* (Wiley, New York, 1976)
17. R.D. Luce, R. Raiffa, *Games and Decisions: Introduction and Critical Survey* (Dover, New York, 1989)
18. H. Raiffa, *Decision Analysis* (Addison-Wesley, Reading, Massachusetts, 1970)
19. C. Kiekintveld, M. Jain, J. Tsai, J. Pita, F. Ordonez, M. Tambe, Computing optimal randomized resource allocations for massive security games, in *Proceedings of the 8th International Joint Conference on Autonomous Agents and Multiagent Systems AAMAS'09*, Budapest, Hungary, 10–15 May 2009
20. T.H. Cormen, C.E. Leiserson, R.L. Rivest, C. Stein, *Introduction to Algorithms* (MIT Press, Cambridge, Massachusetts, 2009)
21. G. Klir, B. Yuan, *Fuzzy Sets and Fuzzy Logic: Theory and Applications* (Prentice Hall, Upper Saddle River, New Jersey, 1995)
22. H.T. Nguyen, E.A. Walker, *First Course on Fuzzy Logic* (CRC Press, Boca Raton, Florida, 2006)
23. O. Lerma, V. Kreinovich, C. Kiekintveld, Linear-time resource allocation in security games with identical fully protective resources, in *Proceedings of the AAAI Workshop on Applied Adversarial Reasoning and Risk Modeling AARM'11*, San Francisco, California, 7 Aug 2011
24. T. Sandler, D.G. Arce, Terrorism and game theory, in *Simulation and Gaming*, vol. 34, no. 3 (2003), pp. 319–337
25. V.M. Bier, Choosing what to protect. Risk Analysis **27**(3), 607–620 (2007)
26. T. Alpcan, T. Basar, A game theoretic approach to decision and analysis in network intrusion detection, in *Proceedings of the 42nd IEEE Conference on Decision and Control CDC'2003*, Maui, Hawaii, Maui, 9–12 Dec 2003, pp. 2595–2600

27. K.C. Nguyen, T.A.T. Basar, Security games with incomplete information, in *Proceedings of IEEE International Conference on Communications ICC'09*, Dresden, Germany, 14–18 June 2009
28. V. Srivastava, J. Neel, A.-B. MacKenzie, R. Menon, L.A. Dasilva, J.E. Hicks, J.H. Reed, R.P. Gilles, Using game theory to analyze wireless ad hoc networks. IEEE Commun. Surv. Tutor. **7**(4), 46–56 (2005)
29. N. Gatti, Game theoretical insights in strategic patrolling: model and algorithm in normal-form, in *Proceedings of the 18th European Conference on Artificial Intelligence ECAI'08*, Patras, Greece, 21–25 July 2008, pp. 403–407
30. N. Agmon, S. Kraus, G.A. Kaminka, V. Sadov, Adversarial uncertainty in multi-robot patrol, in *Proceedings of the 21st International Joint Conference on Artificial Intelligence IJCAI'09*, Pasadena, California, 11–17 July 2009
31. E. Halvorson, V. Conitzer, R. Parr, Multi-step multi-sensor hider-seeker games, in *Proceedings of the 21st International Joint Conference on Artificial Intelligence IJCAI'09*, Pasadena, California, 11–17 July 2009
32. T. Roughgarden, Stackelberg scheduling strategies. SIAM J. Comput. **33**(2), 332–350 (2004)
33. J. Pita, M. Jain, C. Western, C. Portway, M. Tambe, F. Ordonez, S. Kraus, P. Parachuri, Deployed ARMOR protection: the application of a game-theoretic model for security at the Los Angeles International Airport, in *Proceedings of the 7th International Joint Conference on Autonomous Agents and Multiagent Systems AAMAS'08*, Estoril, Portugal, 12–16 May 2008
34. J. Tsai, S. Rathi, C. Kiekintveld, F. Ordóñez, M. Tambe, IRIS—A tool for strategic security allocation in transportation networks, in *Proceedings of the Eighth International Conference on Autonomous Agents and Multiagent Systems AAMAS'09*, Budapest, Hungary, 10–15 May 2009
35. J. Pita, M. Tambe, C. Kiekintveld, S. Cullen, E. Steigerwald, Guards—game theoretic security allocation on a national scale, in *Proceedings of the 10th International Conference on Autonomous Agents and Multiagent Systems AAMAS'11*, Taipei, Taiwan, 2–6 May 2011
36. V. Conitzer, T. Sandholm, Computing the optimal strategy to commit to, in *Proceedings of the 7th ACM Conference on Electronic Commerce EC'06*, Ann Arbor, Michigan, 11–15 June 2006, pp. 82–90
37. P. Paruchuri, J.P. Pearce, J. Marecki, M. Tambe, F. Ordonez, S. Kraus, Playing games with security: an efficient exact algorithm for Bayesian Stackelberg games, in *Proceedings of the 7th International Joint Conference on Autonomous Agents and Multiagent Systems AAMAS'08*, Estoril, Portugal, 12–16 May 2008, pp. 895–902
38. M. Jain, E. Kardes, C. Kiekintveld, M. Tambe, F. Ordonez, Security games with arbitrary schedules: a branch and price approach, in *Proceedings of the Twenty-Fourth AAAI Conference on Artificial Intelligence AAAI'10*, Atlanta, Georgia, 11–15 July 2010
39. H. von Stackelberg, *Marktform und Gleichgewicht* (Springer Verlag, Vienna, 1934)
40. G. Leitmann, On generalized Stackelberg strategies. Optim. Theory Appl. **26**(4), 637–643 (1978)
41. T. Basar, G.J. Olsder, *Dynamic Noncooperative Game Theory* (Academic Press, San Diego, California, 1995)
42. B. von Stengel, S. Zamir, *Leadership with commitment to mixed strategies*, London School of Economics LSE, Computational, Discrete, and Applied Mathematics (CDAM) Report Series, Technical Report LSE-CDAM-2004-01 (2004)

Chapter 3
Data and Knowledge Processing

Abstract Once the data is collected, we need to process this data. For processing, we need to use computers—and the more data we collect, the more computer power we need. It is therefore important to optimally distribute this computing power. This is the problem that we will analyze in Sect. 3.1. In many cases, data processing is a creative process, it goes beyond simple application of known algorithms. The problem of selecting the best ways of organizing this process is handled in Sect. 3.2. To come up with the best ways of processing data, of extracting knowledge from the data, it is often not sufficient to have *individual* efforts, we need creative *teams*—teams that combine domain expertise and computer expertise. Just like with computers, simply bringing people together does not always improve their efficiency. It is therefore important to make sure that people collaborate in the most efficient way. This aspect of data and knowledge processing is analyzed in Sect. 3.3.

3.1 Data and Knowledge Processing: How to Best Organize Computing Power

In this section, we analyze how to best organize the computing power. The results from this section were first published in [1].

Towards the most efficient way of organizing computing power: enter cloud computing. In many application areas (bioinformatics, geosciences, etc.) we need to process large amounts of data, which requires fast computers and fast communication. Historically, there have been limits on the amount of the information that can be transmitted at a high speed, and these limits affected information processing.

A few decades ago, we could only send the results of data processing fast. As a result, the best strategy to speed up computations was to move all the data into a central location, close to the high performance computers for processing this data.

L.O. Lerma and V. Kreinovich, *Towards Analytical Techniques for Optimizing Knowledge Acquisition, Processing, Propagation, and Use in Cyberinfrastructure and Big Data*, Studies in Big Data 29, DOI 10.1007/978-3-319-61349-9_3

In the last decades, it became equally fast to move big portions of databases needed to answer a certain query. This enabled the users to switch to a *cyberinfrastructure* paradigm, when there is no longer need for time-consuming moving of data to a central location: the data is stored where it was generated, and when needed, the corresponding data is moved to processing computers; see, e.g., [2–6] and references therein.

Nowadays, moving the whole databases becomes almost as fast, so there is no longer need to store the data where it was produced—it is possible to store the data where it will be best for future data processing. This idea underlies the paradigm of *cloud computing*. This paradigm is especially important for processing big data, [7–10] when the amount of data is so huge that the individual user does not have enough computation power to process this data and it is, thus, important to borrow additional computation power from the cloud.

What is the most efficient way of cloud computing. The main advantage of cloud computing is that we can make computations more efficient by finding optimal placement of the servers that store and/or process the corresponding data. So, in developing cloud computing schemes, it is important to be able to solve this optimization problem. In this chapter, we consider the corresponding problem of optimal data storage in cloud computing.

Comment. This server placement problem is very similar to the type of problems faced by Akamai and other companies that do web acceleration via caching; we therefore hope that our solution can be of help in web acceleration as well.

Towards a precise formulation of the problem: first approximation. We usually *know* the geographic density $\rho_u(x)$ describing possible users of this particular database (e.g., a database containing geophysical data), and we know the number of duplicates D that we can afford to store. We *need to determine* the storage density $\rho_s(x)$, i.e., number of copies per geographic region, so as to minimize the average communication delay.

First approximation model: main assumption. In the first approximation, we can measure the travel delay by the average travel distance.

Derivation of the corresponding model. How can we describe this distance in terms of the density $\rho_s(x)$? When the density is constant, we want to place the servers in such a way that the largest distance r to a sensor is as small as possible. (Alternatively, if r is fixed, we want to minimize the number of servers for which every point is at a distance $\leq r$ from one of the servers. In geometric terms, this means that every point on a plane belongs to a circle of radius r centered on one the sensors—and thus, the whole plane is covered by such circles. Out of all such coverings, we want to find the covering with the smallest possible number of sensors.

As we have mentioned in Chap. 2, it is known that the smallest such number is provided by an equilateral triangle grid, i.e., a grid formed by equilateral triangles; see, e.g., [11, 12].

Let us assume that we have already selected the server density function $\rho_s(x)$. Within a small region of area A, we have $A \cdot \rho_s(x)$ servers. Thus, if we, e.g., place these servers on a grid with distance h between the two neighboring ones in each direction, we have:

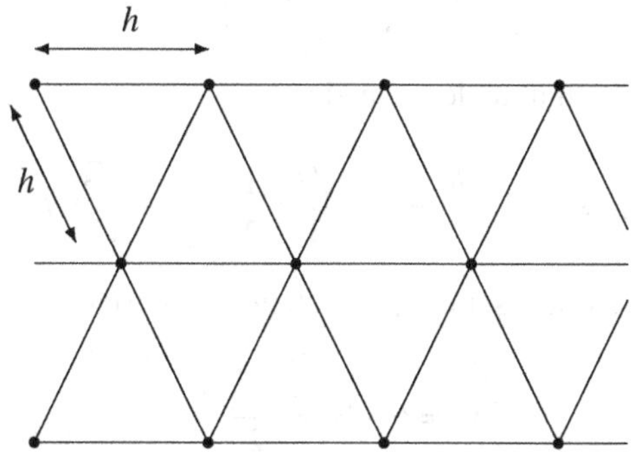

For this placement, the set of all the points which are closest to a given server forms a hexagonal area:

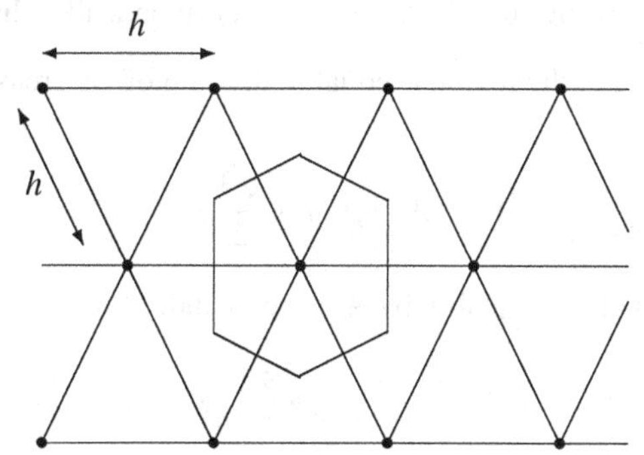

This hexagonal area consists of 6 equilateral triangles with height $h/2$:

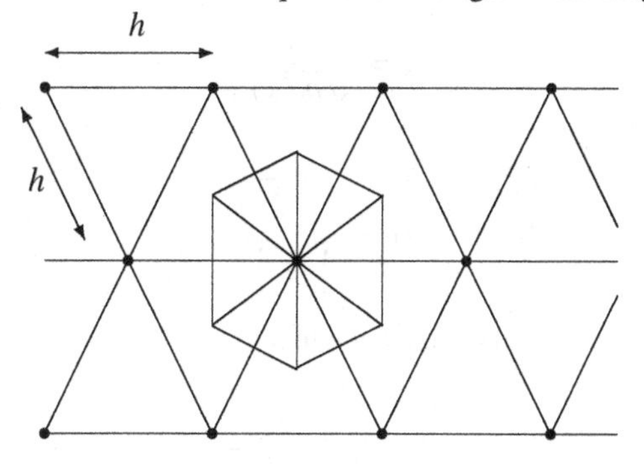

In each triangle, the height $h/2$ is related to the size s by the formula

$$\frac{h}{2} = s \cdot \cos(60^\circ) = s \cdot \frac{\sqrt{3}}{2},$$

hence

$$s = \frac{h}{\sqrt{3}} = h \cdot \frac{\sqrt{3}}{3}.$$

Thus, the area A_t of each triangle is equal to

$$A_t = \frac{1}{2} \cdot s \cdot \frac{h}{2} = \frac{1}{2} \cdot \frac{\sqrt{3}}{3} \cdot \frac{1}{2} \cdot h^2 = \frac{\sqrt{3}}{12} \cdot h^2.$$

So, the area A_s of the whole set is equal to 6 times the triangle area:

$$A_s = 6 \cdot A_t = \frac{\sqrt{3}}{2} \cdot h^2.$$

Each point from the region is the closest to one of the points from the server grid, so the region of area A is thus divided into $A \cdot \rho_s(x)$ (practically) disjoint sets of area $\frac{\sqrt{3}}{2} \cdot h^2$. So, the area of the region is equal to the sum of the areas of these sets:

$$A = (A \cdot \rho_s(x)) \cdot \frac{\sqrt{3}}{2} \cdot h^2.$$

Dividing both sides of this equality by A, we conclude that

$$1 = \rho_s(x) \cdot \frac{\sqrt{3}}{2} \cdot h^2,$$

and hence, that

$$h = \frac{c_0}{\sqrt{\rho_s(x)}},$$

where we denote

$$c_0 \stackrel{\text{def}}{=} \sqrt{\frac{2}{\sqrt{3}}}.$$

The largest distance r to a server is thus equal to

$$\frac{h}{2} = \frac{c_0}{2 \cdot \sqrt{\rho_s(x)}}.$$

The average distance $\overline{\rho}$ is proportional to r—since when we re-scale the picture, all the distances—including the average distance—increase proportionally. Since the distance r is proportional to $(\rho_s(x))^{-1/2}$, the average distance near the location x is thus also proportional to this same value: $\overline{\rho}(x) = \text{const} \cdot (\rho_s(x))^{-1/2}$ for some constant.

At each location x, we have $\sim\rho_u(x)$ users. Thus, the total average distance—the value that we would like to minimize—is equal to $\int \overline{\rho}(x) \cdot \rho_u(x)\,dx$ and is, thus, proportional to

$$\int (\rho_s(x))^{-1/2} \cdot \rho_u(x)\,dx.$$

So, minimizing the average distance is equivalent to minimizing the value of the above integral.

We want to find the server placement $\rho_s(x)$ that minimizes this integral under the constraint that the total number of server is D, i.e., that $\int \rho_s(x) = D$.

Resulting constraint optimization problem. Thus, we arrive at the following optimization problem:

- We know the density $\rho_u(x)$ and an integer D;
- under all possible functions $\rho_s(x)$ for which $\int \rho_s(x)\,dx = D$, we must find a function that minimizes the integral $\int (\rho_s(x))^{-1/2} \cdot \rho_u(x)\,dx$.

Solving the constraint optimization problem. A standard way to solve a constraint optimization problem of optimizing a function $f(X)$ under the constraint $g(X) = 0$ is to use the Lagrange multiplier method, i.e., to apply unconstrained optimization to an auxiliary function $f(X) + \lambda \cdot g(X)$, where the parameter λ (called *Lagrange multiplier*) is selected in such a way so as to satisfy the constraint $g(X) = 0$.

With respect to our constraint optimization problem, this means that we need to select a density $\rho_s(x)$ that optimizes the following auxiliary expression:

$$\int (\rho_s(x))^{-1/2} \cdot \rho_u(x)\,dx + \lambda \cdot \left(\int \rho_s(x)\,dx - D \right).$$

Having an unknown function $\rho_s(x)$ means, in effect, that we have infinitely many unknown values $\rho(x)$ corresponding to different locations x. Optimum is attained when the derivative with respect to each variable is equal to 0. Differentiating the above expression with respect to each variable $\rho_s(x)$, and equating the result to 0, we get the equation

$$-\frac{1}{2} \cdot (\rho_s(x))^{-3/2} \cdot \rho_u(x) + \lambda = 0,$$

hence $\rho_s(x) = c \cdot (\rho_u(x))^{2/3}$ for some constant c.

The constant c can be determined from the constraint $\int \rho_s(x)\,dx = D$, i.e., that

$$\int c \cdot (\rho_u(x))^{2/3}\,dx = c \cdot \int (\rho_u(x))^{2/3}\,dx = D.$$

Thus,

$$c = \frac{D}{\int (\rho_u(x))^{2/3}\,dx},$$

and we arrive at the following solution.

Solution to the problem. Once we know the user density $\rho_u(x)$ and the total number of servers D that we can afford, the optimal server density $\rho_s(x)$ is equal to

$$\rho_s(x) = D \cdot \frac{(\rho_u(x))^{2/3}}{\int (\rho_u(y))^{2/3}\,dy}.$$

Discussion. In line with common sense, the optimal server density increases when the user density increases, i.e.:

- in locations where there are more users, we place more servers, and
- in locations where there are fewer users, we place fewer servers.

However, when the user density decreases, the server density decreases slower—because otherwise, if we took the server density simply proportional to the user density, the delays in areas with few users would have been huge.

Comment. From the mathematical viewpoint, this analysis is similar to the analysis of a security-related optimization problem, in which, instead of placing servers, we need to place sensors; see [12].

Towards a more realistic model: first idea. In the above first approximation, we only took into account the time that it takes to move the data to the user. This would be all if the database was not changing. In real life, databases need to be periodically updated. Updating also takes time. Thus, when we find the optimal placement of servers, we need to take into account not only expenses on moving the data to the users, but also the expenses of updating the information.

Towards a precise formulation of this idea. How do we estimate these expenses? In a small area, where the user distribution is approximately uniform, the servers are also uniformly distributed, i.e., they form a grid with distance $h = 2r$ between the two neighboring servers [11, 12]. Within a unit area, there are $\sim 1/r^2$ servers, and reaching each of them from one of its neighbors requires time proportional to the distance $\sim r$. The overall effort of updating all the servers can be obtained by multiplying the number of servers by an effort needed to update each server, and is thus proportional to $1/r^2 \cdot r \sim 1/r$. We already know that $r \sim (\rho_s(x))^{-1/2}$, thus, the cost of updating all the servers in the vicinity of a location x is proportional to $(\rho_s(x))^{1/2}$. The overall update cost can thus be obtained by integrating this value over the whole area. Thus, we arrive at the following problem.

Resulting optimization problem:

- We know the density $\rho_u(x)$, an integer D, and a constant C that is determined by the relative frequency of updates in comparison with frequency of normal use of the database;

- under all possible functions $\rho_s(x)$ for which $\int \rho_s(x)\,dx = D$, we must find a function that minimizes the expression

$$\int (\rho_s(x))^{-1/2} \cdot \rho_u(x)\,dx + \int C \cdot (\rho_s(x))^{1/2}\,dx.$$

Solving the problem. To solve the new optimization problem, we can similarly form the Lagrange multiplier expression

$$\int (\rho_s(x))^{-1/2} \cdot \rho_u(x)\,dx + \int C \cdot (\rho_s(x))^{1/2}\,dx + \lambda \cdot \left(\int \rho_s(x)\,dx - D\right),$$

differentiate it with respect to each unknown $\rho_s(x)$, and equate the resulting derivative to 0. As a result, we get an equation

$$-\frac{1}{2} \cdot (\rho_s(x))^{-3/2} \cdot \rho_u(x) + \frac{1}{2} \cdot C \cdot (\rho_s(x))^{-1/2} + \lambda = 0.$$

This is a cubic equation in terms of $(\rho_s(x))^{-1/2}$, so while it is easy to solve numerically, there is no simple analytical expression as in the first approximation case.

The resulting solution $\rho_s(x)$ depends on the choice of the Lagrange multiplier λ, i.e., in effect, we have $\rho_s(x) = \rho_s(x, \lambda)$. The value λ can be determined from the condition that $\int \rho_s(x, \lambda)\,dx = D$.

Second idea. The second idea is that usually, a service provides a time guarantee, so we should require that no matter where a user is located, the time for this user to get the desired information from the database should not exceed a certain value. In our model, this means that a distance r from the user to the nearest server should not exceed a certain given value r_0. Since $r \sim (\rho_s(x))^{-1/2}$, this means, in turn, that the server density should not decrease below a certain threshold ρ_0.

This is an additional constraint that we impose on $\rho_s(x)$. In the first approximation model, it means that instead of the formula $\rho_s(x) = c \cdot (\rho_u(x))^{2/3}$—which could potentially lead to server densities below ρ_0—we should have $\rho_s(x) = \max(c \cdot (\rho_u(x))^{2/3}, \rho_0)$.

The parameter c can be determined from the constraint

$$\int \rho_s(x)\,dx = \int \max(c \cdot (\rho_u(x))^{2/3}, \rho_0)\,dx = D.$$

Since the integral is an increasing function of c, we can easily find the solution c of this equation by bisection (see, e.g., [13]).

Combining both ideas. If we take both ideas into account, then we need to consider only those roots of the above cubic equation which are larger than or equal to ρ_0; if all the roots are $< \rho_0$, we take $\rho_s(x) = \rho$.

The resulting solution $\rho_s(x)$ depends on the choice of the Lagrange multiplier λ, i.e., in effect, we have $\rho_s(x) = \rho_s(x, \lambda)$. The corresponding value λ can also be similarly determined from the equation $\int \rho_s(x, \lambda)\, dx = D$.

3.2 Data and Knowledge Processing: How to Best Organize Research

In a book aimed at young researchers, a renowned biologist E. O. Wilson advises them to be persistent, to continuously follow the same research direction. While this advice is supported by his (and others') experience of mentoring young scientists, it seems to contradict the experience of many famous scientists of the past who moved from one research direction to another and still achieved great success. In this section, we provide a geometric explanation for Wilson's advice—and we explain why this explanation was not applicable in the past.

Results from this section first appeared in [14].

Sustained effort, persistence are important: an advice. In his book [15], the renowned biologist E. O. Wilson advises to young scientists on how to do research. One of his main pieces of advice is the need for sustained effort, the need to persist, to keep going despite obstacles, losses or setbacks, to continue moving forward until one have completed his/her goals.

Why this advice? On the one hand, this advice makes sense, and it is supported by the experience of mentoring young researchers.

On the other hand, this advice seems contrary to the general spirit of creative research: instead of continuing to go in the same direction, why not try different things?

The history of science seems to support the advantages of going in different directions, this is how most researchers succeeded in the past: Newton invented calculus, discovered Newton'slaws of mechanics, studied optics, and make comments on the Bible. Why not advise young researchers to follow Newton's example?

What we do in this section. In this section, we use a simple geometric model of expanding knowledge to show that—while creativity is important—the optimal scientific strategy is indeed to be persistent.

Expanding knowledge: a natural description. We as a civilization have started with the basic knowledge, and we have been expanding this knowledge in all directions. At each moment of time, we have expanded our knowledge only so far, and we continue expanding as time goes on.

It is therefore reasonable to characterize our state of knowledge at each moment of time by the distance R from the basic knowledge that we have reached by this moment.

The resulting geometric description. As a result, we arrive at the following description of our current state of knowledge:

- the set of all possible pieces of knowledge is a multi-dimensional space;
- in this space, there is a fixed point O that represents the basic knowledge;
- at each moment of time t, the civilization's state knowledge is represented by a number R; crudely speaking, we know all the pieces of knowledge whose distance from the point O does not exceed R.

The value R increases with time, meaning that, as we learn new things, the radius R of our "sphere of knowledge" increases.

This is the model that we will use in our analysis.

The main objective of research reformulated in terms of the geometric model. Our goal is to expand the knowledge, i.e., increase the "radius of knowledge" R as much as possible.

For that purpose, numerous scientists are positioned at the edge of this sphere of knowledge. Each of these scientists moves ahead into the unknown territory. As a result, the sphere of knowledge expands.

Limitations. As a scientist progresses, he or she moves from the original sphere of science into the unknown. The rate with which a scientist can uncover new knowledge is limited. In geometric terms, this means that there is a bound B on the speed with which a scientist can move.

Because of this bound, the farthest a scientist can move during the time period T is the distance $B \cdot T$.

In which direction should a scientist go? Our goal is to expand the sphere of knowledge as much as possible. With researchers positioned all over the current border of knowledge (i.e., at the distance R from the point O), the expansion is determined by how far each researcher will go.

If each researcher who is originally at a point S on the sphere of radius R continues along the straight line OS past the point S, then after the period T, this researcher will reach a new point S', at the distance $R' \stackrel{\text{def}}{=} R + B \cdot T$ from the point O. If all researchers follow this strategy, our radius of knowledge will increase from R to $R' = R + B \cdot T$.

Let us show that this "following the straight line" idea is the optimal research strategy. Indeed, it is well known that the straight line is the shortest of all the paths connecting two points. Thus, if a researcher's trajectory of the overall length $B \cdot T$ is not a straight line, then the distance between the starting point S and the final point S' is smaller than $B \cdot T$: $d(S, S') < B \cdot T$. As a result, due to the triangle inequality $d(O, S') \le d(O, S) + d(S, S')$, we get $d(O, S') < R + B \cdot T$. In other words, in this case, the sphere of knowledge expands less that in situation where each researcher follows a straight line.

We arrive at the desired explanation. We therefore conclude that an optimal way to increase knowledge is for each scientist to follow a straight line, to go in the same direction all the time—i.e., to be persistent. Thus, *a natural geometric model of* *expanding knowledge indeed provides an explanation of why persistence (= sustained effort) is important.*

What about Newton? So why did Newton (and many other famous scientists of the past) not follow this strategy? Why are our arguments not applicable to Newton?

The main difference between now and Newton's times is that in the past, we did not have researchers positioned all over the border of knowledge. The reason for this difference is simple: in the past, there were much fewer researchers. For example, in the time of Newton, in addition to Newton himself, there was only one other person in the world working on calculus—Leibniz.

As a result, in the past, to expand knowledge in all areas, the same researcher had to enhance the knowledge's expansion in several areas—thus deviating from the straight line trajectory.

Nowadays, in each area of research, no matter how narrow, there are dozens of researchers. As a result, the most efficient way to collectively expand knowledge is for each of them to be persistent—and also to team up with other researchers. This brings us to the topic of the next section.

3.3 Data and Knowledge Processing: How to Best Organize Research Teams

How to best organize research teams? To solve this problem, we propose a natural model describing competition between two research groups of the same average research strength. The analysis of this model enables us to conclude that a more diverse group has an advantage: namely, the more diverse the group, the higher the average quality of its publications.

Of course, this should not be take to an extreme: if the difference between participants in a team is too large, they will not be able to successfully collaborate. However, up to that level of difference, an increase in diversity is beneficial.

3.3.1 Diversity Is Important: Main Argument

In this section, we explain that in the organization of the research teams, it is important to maintain diversity. The results from this section were first published in [16].

Diversity is beneficial. Experiments and simulation have shown that, in general, more diverse groups have an advantage over less diverse ones; see, e.g., [17–20].

What we do in this section. In this section, we provide an additional quantitative argument in favor of diversity of research groups. Namely, we show that if we have two competing research groups with the same average strength, then the more diverse research group has a clear advantage.

Natural assumption: strength is normally distributed. Normal distributions are ubiquitous, they appear in many real-life situations; in particular, they describe the distribution of many characteristics of a human being such as height, weight, blood pressure, or IQ. The ubiquity of normal distribution can be explained by the fact that in many cases, the value of the quantity is caused by many independent factors, and the known Central Limit Theorem states, crudely speaking, that the distribution of the sum of large number small independent factors is close to normal; see, e.g., [21].

It is therefore reasonable to assume that within each of the two competing research groups, strength is normally distributed. In general, a normal distribution is uniquely determined by its mean μ and its standard deviation σ. In terms of strength, the mean is the average strength, while the standard deviation describe diversity: the larger the standard deviation, the more diverse the group.

We assume that both groups have the same average strength μ, but that the first group is more diverse: $\sigma_1 > \sigma_2$.

How the groups compete: a description. We assume that each group coordinate the research efforts of its members, so there is no unnecessary competition within each group; the only competition is between the two groups. Once a member of one of the research groups selects a problem—a problem that people in the field consider to be important—it is highly probably that the same problem will be picked up by some member of another research group.

The groups (being competitors) do not coordinate their research efforts with each other. As a result, the corresponding member of another research group is randomly selected from that group. If two researchers of different research strength $s_1 > s_2$ work on the same problem, it is reasonable to expect that the stronger researcher will get the results first—and this will result in a publication of quality corresponding to this higher strength s_1.

Let us analyze the resulting model.

Analysis of the model. Under the above assumptions, let us see which of the two groups has an advantage. Intuitively, the answer is not clear:

- on the one hand, the more diverse research group has a larger number of stronger researchers, which gives this group an *advantage* over the less diverse group;
- on the other hand, the more diverse research group also has a larger number of weaker researchers, which gives this group a *disadvantage* over the less diverse group.

At first glance, diversity brings no advantage. In the above competition, which of the two groups will be more successful? Let us first consider the simplest measure of success: the resulting number of publications.

The first group gets a publication is a value s_1 randomly selected from the first group exceeds a value s_2 randomly selected from the second group: $s_1 > s_2$. Thus, the number of publications produced by the first group is proportional to the probability that for randomly selected values s_1 and s_2, we have $s_1 - s_2 > 0$, i.e., equivalently, that $s_1 - s_2 > 0$. The two independent random variables x_1 and x_2 are normally distributed with the same mean μ. It is known that the difference of two independent normally distributed random variables is also normally distributed. The mean of the difference $s_1 - s_2$ is equal to the difference of the means, i.e., to $\mu - \mu = 0$. Thus, $s_1 - s_2$ is a normally distributed random variable with 0 mean. For such random variable, the probability of it being positive is exactly $\frac{1}{2}$.

Thus, when the two research groups have the same average strength, in half of the cases, the first group will succeed, in half of the cases, the second group will succeed. So, both groups will generate, on average, the same number of publications.

Towards a deeper analysis. In terms of *number* of published papers, diversity does not bring any advantage. However, different publications have different *quality*. What if, instead of simply counting the number of publications, we would instead estimate the average quality of a publication?

According to our model, the first group succeeds if $s_1 > s_2$ and produces a paper of quality s_1. Thus, the average quality q_1 of papers produced by the first research group is equal to the conditional expectation $q_1 = E[s_1 \mid s_1 > s_2]$. Similarly, the average quality q_2 of papers produced by the second research group is equal to the conditional expectation $q_2 = E[s_2 \mid s_1 > s_2]$.

Let us estimate these two quantities.

Estimating the desired quantities. The first research group produces a paper of quality s_1:

- if there is a person of strength s_1 in this group and
- if this person was stronger than the competitor, i.e., a person with a (randomly selected) strength s_2 from the second research group.

The probability of the first research group having a member of strength s_1 is determined by the normal distribution, i.e., has the form

$$f_1(s_1) = \frac{1}{\sigma_1} \cdot \phi\left(\frac{s_1 - \mu}{\sigma_1}\right),$$

where

$$\phi(x) \stackrel{\text{def}}{=} \frac{1}{\sqrt{2\pi}} \cdot \exp\left(-\frac{x^2}{2}\right)$$

is the probability density of the standard normal distribution (i.e., a normal distribution with mean 0 and standard deviation 1).

The probability that s_1 will win over the competitor is equal to

$$\text{Prob}(s_2 < s_1).$$

By definition of the cumulative distribution function (cdf) $F_2(x)$ of the random variable s_2, this probability is equal to $\text{Prob}(s_2 < s_1) = F_2(s_1)$. Since the variable s_2

is normally distributed, this probability has the form $F_2(s_1) = \Phi\left(\frac{s_1 - \mu}{\sigma_2}\right)$, where $\Phi(x)$ is the cdf of the standard normal distribution.

Since s_1 and s_2 are independent, the probability distribution function $f(s_1)$ for the publication quality s_1 is proportional to the product of the two probabilities, i.e., has the form

$$f(s_1) = \text{const} \cdot \phi\left(\frac{s_1 - \mu}{\sigma_1}\right) \cdot \Phi\left(\frac{s_1 - \mu}{\sigma_2}\right).$$

Such a distribution is known: it is a *skew-normal* distribution; see, e.g., [22–24] and references therein. To be more precise, the usual formula for the skew-normal distribution has the form

$$f(s_1) = \text{const} \cdot \phi\left(\frac{s_1 - \mu}{\sigma_1}\right) \cdot \Phi\left(\alpha \cdot \left(\frac{s_1 - \mu}{\sigma_1}\right)\right),$$

which coincides with the above form for $\alpha = \frac{\sigma_1}{\sigma_2}$.

It is known that the mean value of the skew-normal random variable is equal to $q_1 = \mu + \sqrt{\frac{2}{\pi}} \cdot \sigma_1 \cdot \frac{\alpha}{\sqrt{1 + \alpha^2}}$. Substituting $\alpha = \frac{\sigma_1}{\sigma_2}$ into this formula and multiplying both the numerator and the denominator of the corresponding fraction by σ_2, we conclude that

$$q_1 = \mu + \frac{\sigma_1^2}{\sqrt{\sigma_1^2 + \sigma_2^2}}.$$

Similarly, the average quality of papers published by the second research group is equal to

$$q_2 = \mu + \frac{\sigma_2^2}{\sqrt{\sigma_1^2 + \sigma_2^2}}.$$

Preliminary summary of the results. From the above formulas, we can see that the larger the standard deviation σ_i, the larger the average quality q_i of the corresponding publications. Thus, while a diverse group produces, on average, the same number of publications, the average quality of these publications is higher—and the more diverse the group, the higher the quality.

3.3.2 *Diversity Is Important: Additional Argument*

In this section, we show that mathematical results from the theory of deductive systems, results which have been used to explain the evolutionary advantage of sexual reproduction over asexual one, can be also used to explain potential advantages of interdisciplinary research and of diversity in the workplace.

Results of this section first appeared in [25].

Background. A 1987 MIT book [26] by a renowned logician S. Yu. Maslov contains, among other things, a possible mathematical explanation of why sexual reproduction is more efficient than asexualn. This explanation is based on the analysis of deductive systems.

When the environment changes, the original DNA—which was adequate for survival in the previous environment—is often no longer adequate. In this case, for the species to survive, they need to modify their DNA so as to make it more adequate for the changed environment.

In general, there are main ways to change individual DNAs:

- via *asexual reproduction*, in which the parent's genes sequence pass to the offspring, usually with a mutation, and
- via *sexual reproduction*, in which the parents' genes are mixed (recombined) and a mutation is added to form the offspring's gene sequence.

The book [26] compares the smallest number of generations that are needed, for both ways, to achieve the desired change. The book proves a mathematical theorem, according to which, under some reasonable assumptions, the possibility of recombination makes the change exponentially faster.

This result explains the evolutionary advantage of sexual reproduction in precise mathematical terms.

Comment. The book [26] also gives an interesting explanation of why only two sexes are used and not three or four: namely, it proves that, in general, adding extra sexes will not speed up the process any further.

Inter-disciplinary research versus research within a discipline: idea.

- When we work within a single discipline, it is more like mutations (while not necessarily a random one).
- On the other hand, inter-disciplinary research provides an opportunity to combine techniques and results from different disciplines.

In view of this analogy, the speed-up result from [26] explains the advantages of inter-disciplinary research.

Comment. Similar ideas appeared in [27].

Diversity in workplace: idea. A similar idea can explain the advantages of diversity in workplace, where:

- working within a single culture is similar to asexual reproduction, while
- diversity provides us with the possibility to productively combine several different cultural viewpoints.

Thus, the main result from [26] explains potential drastic advantages of diversity.

Comment. Similar arguments explaining the benefits of diversity have also been described in [17–20, 28–30].

3.3.3 Caution: Excessive Diversity May Be Detrimental

What if diversity is too high? As we have mentioned earlier, experiments show that when we place people of different skill level in a team, this diversity enhances the productivity of a team. For example, when graduate students and senior undergraduate students work together on a joint research project, it helps students from both groups. Undergraduate students learn a lot from the graduate students. On the other hand, when the undergraduate students start asking detailed questions, this helps the graduate students better understand which parts of the material they mastered already and in which they still need to learn more.

On the other hand, if we bring together graduate students and middle school students, then—unless the graduate students have pedagogical talent—we will not get a successful collaboration. How do we capture this difference in the above model?

A seeming contradiction with our analysis? At first glance, our analysis seems to indicate that no matter how high the difference, the team with a larger degree of diversity is always more successful. However, a more detailed analysis of this situation shows that this is not the case.

Re-analyzing the model. Indeed, in our analysis, we made an assumption that in each team, the distribution of strength is normal. In both teams, the average strength is the same; we denoted this joint average strength by μ. Without losing generality, let us assume that the first team is more diverse, i.e., the standard deviation σ_1 of the first team is larger the standard deviation σ_2 of the second team.

It is known that for the normal distribution with mean μ and standard deviation σ, with probability 99.9% the value is located in the three-sigma interval

$$[\mu - 3\sigma, \mu + 3\sigma].$$

For a research team, this means that only 0.1% of its members—i.e., one on a thousand—have strength values which are outside this interval. Since the number of people in a research team is usually much smaller than a thousand, this means that the strengths of all the members in a research team is within the corresponding three-sigma interval $[\mu - 3\sigma_i, \mu + 3\sigma_i]$.

With probability 95%, each value of a normally distributed random variable is located in a two-sigma interval $[\mu - 2\sigma, \mu + 2\sigma]$. So, for research groups with less than 20 participants, the strengths of all the members in a research team is located within the corresponding narrower two-sigma interval $[\mu - 2\sigma_i, \mu + 2\sigma_i]$. In general, depending on the size of the research team, we can claim that the strengths of all the members of this team is within the interval $[\mu - k_0 \cdot \sigma_i, \mu + k_0 \cdot \sigma_i]$, for an appropriate value k_0. For appropriate k_0, in each group, we have researchers whose strength s is close to $\mu - k_0 \cdot \sigma_i$, and we also have researchers whose strength s is close to $\mu + k_0 \cdot \sigma_i$.

Since $\sigma_1 > \sigma_2$, the interval $[\mu - k_0 \cdot \sigma_1, \mu + k_0 \cdot \sigma_1]$ describing the strengths of the first team is wider than the interval $[\mu - k_0 \cdot \sigma_2, \mu + k_0 \cdot \sigma_2]$ describing the strengths of the researchers from the second team. In particular, this means that the strength $\mu - k_0 \cdot \sigma_1$ of the weakest member of the first team is smaller than the strength $\mu - k_0 \cdot \sigma_2$ of the weakest member of the second team—and is, thus, weaker than the strength of *all* the members from the second team. As a result, no matter what problem this weakest member of the first team will start working on, the member of the second team who starts working on the same problem will get his or her result first. In other words, this weakest member of the first team does not contribute anything to the success of the first team in its competition with the second team—and can, therefore, be dismissed from this first team. After this dismissal, the diversity of the first team decreases—but its average strength increases, and the success rate does not change.

This analysis explains why excessive diversity is detrimental. The above analysis shows that while, in general, diversity is beneficial, if we have a team where the diversity is too high (such as graduate students and middle school students), then the weakest members of this team are not able to productively participate and can, thus, be safely dismissed from the team without changing its productivity.

With this comment in mind, we can now come up with the actual summary of the results.

Summary of the results. Up to a certain level, the larger the standard deviation σ_i, the larger the average quality q_i of the corresponding publications. Thus, while a diverse group produces, on average, the same number of publications, the average quality of these publications is higher—and the more diverse the group, the higher the quality.

However, one needs to be careful not to overdo it: if the diversity is too high, the difference between the team members is too large, and a productive collaboration is no longer possible.

3.3.4 *Towards Finding the Optimal Level of Diversity*

In the previous sections, we mentioned that while, in general, diversity of a research team is beneficial for research collaboration, excessive diversity can be detrimental. It is therefore desirable to develop recommendations for an optimal level of diversity.

Our analysis was based on the assumptions about the distribution of the scientists' strength. In the above sections, we used a first-order approximation, in which we assumed that the strength is normally distributed. In this section, we show how to develop more adequate models of this distributions, models which will hopefully eventually lead to practical recommendations about the optimal level of diversity.

Results from this section first appeared in [16, 31].

Idea. To become a professional scientist, one has to defend his/her PhD. Not all students who start their PhD studies end up with a dissertation: some students succeed, but many don't. Crudely speaking, a student succeeds if his/her strength is sufficient to solve the corresponding problem, i.e., in other words, if his/her strength is larger than or equal to the complexity of the selected problem.

Transforming the above idea into a precise model. It is reasonable to assume that the strength x of students entering a PhD program is normally distributed, with some mean μ and standard deviation σ. (The strength is caused by many different factor, so it is reasonable to apply the Central Limit Theorem.) It is similarly reasonable to assume that the complexity y of a problem is normally distributed, with some mean μ' and some standard deviation σ'.

Because of the above assumptions, the number of students of strength x who enter the PhD program is proportional to $f_0\left(\frac{x-\mu}{\sigma}\right)$. It is also reasonable to assume that a student picks a problem at random. Thus, out of the incoming students of strength x, the proportion of those who succeed is equal to the probability $\text{Prob}_y(y \le x)$ that the randomly selected problem has complexity $\le x$—i.e., to the value $F(x)$ of the corresponding cdf. Since the complexities are normally distributed, this probability is equal to $F_0\left(\frac{x-\mu'}{\sigma'}\right)$.

The resulting number of scientists of strength x can be obtained by multiplying the number $\text{const}\cdot f_0\left(\frac{x-\mu}{\sigma}\right)$ of incoming students of strength x by the proportion $F_0\left(\frac{x-\mu'}{\sigma'}\right)$ of those who successfully get their PhD degrees. Thus, the probability density function that described scientists with PhDs is equal to

$$\text{const}\cdot f_0\left(\frac{x-\mu}{\sigma}\right)\cdot F_0\left(\frac{x-\mu'}{\sigma'}\right).$$

This is exactly the skew-normal distribution!

Towards a more detailed model: idea. In the above analysis, to determine whether a student succeeds or not in solving the corresponding problem, we only took into account the student's strength and the problem's complexity. In practice, often, there is an additional factor affecting the student's success: the presence of competition.

In a well-organized university department, students' topics are distributed in such a way that an unproductive competition between students from the same university be avoided. However, since students from different universities handle largely the same problems, competition between students from different university is inevitable.

If we take this competition into account, then we see that for a student to succeed, it is not enough that this student's strength is larger than or equal to the complexity of the problem, it is also important to make sure that the student solves the problem ahead of the competition, i.e., that his/her strength is larger than than the strengths of students from other departments who select the same problem.

Transforming the above idea into a precise model. For a student of strength x to succeed, this strength must be larger than or equal to the complexity y of the selected problem and also great then the strengths $x_1, \ldots, x_n$ of students from competing universities who handle the same problem. In this case, the probability of a student succeeding is equal to the probability that

$$y \le x \text{ and } x_1 \le x \text{ and } \ldots \text{ and } x_n \le x.$$

It is reasonable to assume that the corresponding distributions are independent, so this probability is equal to the product of the corresponding probabilities

$$\text{Prob}_y(y \le x) \cdot \text{Prob}_1(x_1 \le x) \cdot \ldots \cdot \text{Prob}_n(x_n \le x).$$

For each university i, the strengths x_i are normally distributed with mean μ_i and standard deviation σ_i. Thus, the probability that a student of strength x succeeds is equal to the product

$$F_0\left(\frac{x-\mu'}{\sigma'}\right) \cdot F_0\left(\frac{x-\mu_1}{\sigma_1}\right) \cdot \ldots \cdot F_0\left(\frac{x-\mu_n}{\sigma_n}\right).$$

As a result, the probability density function that describes scientists with PhDs is equal to

$$\text{const} \cdot f_0\left(\frac{x-\mu}{\sigma}\right) \cdot F_0\left(\frac{x-\mu'}{\sigma'}\right) \cdot F_0\left(\frac{x-\mu_1}{\sigma_1}\right) \cdot \ldots \cdot F_0\left(\frac{x-\mu_n}{\sigma_n}\right).$$

This is a generalization of the skew-normal distribution, in which the original pdf is multiplied not by one normal cdf, but possible by many normal cfds.

References

1. O. Lerma, E. Gutierrez, C. Kiekintveld, V. Kreinovich, Towards optimal knowledge processing: from centralization through cyberinsfrastructure to cloud computing. Int. J. Innovative Manage. Inf. Prod. (IJIMIP) **2**(2), 67–72 (2011)
2. A. Gates, V. Kreinovich, L. Longpré, P. Pinheiro da Silva, G.R. Keller, Towards secure cyber-infrastructure for sharing border information, in *Proceedings of the Lineae Terrarum: International Border Conference*, El Paso, Las Cruces, and Cd. Juarez, 27–30 Mar 2006
3. G.R. Keller, T.G. Hildenbrand, R. Kucks, M. Webring, A. Briesacher, K. Rujawitz, A.M. Hittleman, D.J. Roman, D. Winester, R. Aldouri, J. Seeley, J. Rasillo, T. Torres, W.J. Hinze, A. Gates, V. Kreinovich, L. Salayandia, A community effort to construct a gravity database for the United States and an associated Web portal, in *Geoinformatics: Data to Knowledge*, ed. by A.K. Sinha (Geological Society of America Publ, Boulder, Colorado, 2006), pp. 21–34
4. L. Longpré, V. Kreinovich, How to efficiently process uncertainty within a cyberinfrastructure without sacrificing privacy and confidentiality, in *Computational Intelligence in Information Assurance and Security*, ed. by N. Nedjah, A. Abraham, and L. de Macedo Mourelle (Springer, 2007), pp. 155–173
5. P. Pinheiro da Silva, A, Velasco, M. Ceberio, C. Servin, M.G. Averill, N. Del Rio, L. Longpré, V. Kreinovich, Propagation and provenance of probabilistic and interval uncertainty in cyberinfrastructure-related data processing and data fusion, in *Proceedings of the International Workshop on Reliable Engineering Computing REC'08*, ed. by R.L. Muhanna and R.L. Mullen, (Savannah, Georgia, 20–22 Feb 2008), pp. 199–234
6. A.K. Sinha (ed.), *Geoinformatics: Data to Knowledge* (Geological Society of America Publ, Boulder, Colorado, 2006)

7. B. Baesens, *Analytics in a Big Data World: The Essential Guide to Data Science and its Applications* (Wiley, Hoboken, New Jersey, 2014)
8. B. Marr, *Big Data: Using SMART Big Data* (Analytics and Metrics To Make Better Decisions and Improve Performance, Wiley, Chichester, UK, 2015)
9. N. Marz, J. Warren, *Big Data: Principles and Best Practices of Scalable Realtime Data Systems* (Manning Publ, Shelter Island, New York, 2015)
10. V. Mayer-Schönberger, K. Cukier, *Big Data: A Revolution That Will Transform How We Live, Work, and Think* (John Murray Publ., London, UK, 2013)
11. R. Kershner, The number of circles covering a set. Am. J Math. **61**(3), 665–671 (1939)
12. C. Kiekintveld and O. Lerma, "Towards optimal placement of bio-weapon detectors", *Proceedings of the 30th Annual Conference of the North American Fuzzy Information Processing Society NAFIPS'2011*, El Paso, Texas, March 18–20, 2011
13. T.H. Cormen, C.E. Leiserson, R.L. Rivest, C. Stein, *Introduction to Algorithms* (MIT Press, Cambridge, Massachusetts, 2009)
14. L. O. Lerma, O. Kosheleva, Why sustained effort is important in research: a geometric explanation. J. Uncertain Syst. **10**, (2016) to appear
15. E.O. Wilson, *Letters to a Young Scientist* (Liveright, New York, 2013)
16. K. Suriya T. Sudtasan, Fighter ants algorithm with application to efficient team building, in *Abstracts of the 7th International Conference of Thailand Econometric Society TES'2014*, Chiang Mai, Thailand, 8–10 Jan 2014
17. L. Hong, S.E. Page, Problem solving by heterogeneous agents. J. Econ. Theor. **97**(1), 123–163 (2001)
18. L. Hong, S.E. Page, Groups of diverse problem solvers can outperform groups of high-ability problem solvers. Proc. Nat. Acad. Sci. **101**(46), 16385–16389 (2004)
19. J.H. Miller, S.E. Page, *Complex Adaptive Social Systems: The Interest in Between* (Princeton University Press, Princeton, New Jersey, 2006)
20. S.E. Page, *The Difference: How the Power of Diversity Creates Better Groups, Firms, Schools, and Societies* (Princeton University Press, Princeton, New Jersey, 2007)
21. D.J. Sheskin, *Handbook of Parametric and Nonparametric Statistical Procedures* (Chapman & Hall/CRC, Boca Raton, Florida, 2011)
22. A. Azzalini, A. Capitanio, *The Skew-Normal and Related Families* (Cambridge University Press, Cambridge, Massachusetts, 2013)
23. B. Li, D. Shi, and T. Wang, Some applications of one-sided skew distributions. Int. J. Intell. Technol. Appl. Stat. **21**, (2009)
24. T. Wang, B. Li, A.K. Gupta, Distribution of quadratic forms under skew normal settings. J. Multivar. Anal. **100**, 533–545 (2009)
25. L.O. Lerma, O. Kosheleva, One more mathematical argument in favor of interdisciplinary research and diversity. Int. Math. Forum **9**(32), 1611–1614 (2014)
26. S.Yu. Maslov, *Theory of Deductive Systems and Its Applications* (MIT Press, Cambridge, Massachusetts, 1987)
27. P. Pinheiro da Silva, A. Velasco, O. Kosheleva, V. Kreinovich, How AI-type uncertainty ideas can improve inter-disciplinary collaboration and education: lessons from a case study. J. Adv. Comput. Intell. Intell. Inf. JACIII **14**(6), 700–707 (2010)
28. O. Castillo, P. Melin, E. Gamez, V. Kreinovich, and O. Kosheleva, "Intelligence techniques are needed to further enhance the advantage of groups with diversity in problem solving", in *Proceedings of the 2009 IEEE Workshop on Hybrid Intelligent Models and Applications HIMA'2009*, Nashville, Tennessee, March 30 – April 2, 2009, pp. 48–55
29. O. Kosheleva, Diversity is the optimal education strategy: a mathematical proof. Int. J. Innovative Manage. Inf. Prod. (IJIMIP) **4**(1), 1–8 (2013)
30. C.C. Kuo, F. Glover, K.S. Dhir, Analyzing and modeling the maximum diversity problem by zero-one programming. Decis. Sci **24**(6), 1171–1185 (1993)
31. K. Suriya, T. Sudtasan, T. Wang, O. Lerma, V. Kreinovich, A natural simple model of scientists' strength leads to skew-normal distribution. Int. J. Intell.Technol. Appl. Stat. (IJITAS) **8**(2), 153–158 (2015)

Chapter 4
Knowledge Propagation and Resulting Knowledge Enhancement

Abstract Once we have transformed data into knowledge, we need to propagate this knowledge—so that other researchers can use and enhance this knowledge. For that, first, we need to motivate people to learn the new knowledge, we need to make sure that the idea is propagated to more and more people. To ensure that, we need to analyze the process of idea propagation; this is done in Sect. 4.1. Once a person is willing to learn the corresponding techniques and ideas, we can start the actual learning. For this learning to be successful, we need to get a good understanding of where the person stands now, what is his/her level of knowledge in the corresponding areas. This assessment problem is analyzed in Sect. 4.2. Once this information is known, we need to actually present this information to the interested folks—and use appropriate feedback to modify (if needed) the speed with which this knowledge is presented. Issues related to the material's presentation are analyzed in Sects. 4.3 and 4.4. Specifically, in Sect. 4.3, we consider the problem from the global viewpoint: e.g., in what order we should present different parts of the material, and how much flexibility should we give to students. In Sect. 4.4, we consider this problem from the local viewpoint: what is the best way to present different items. Finally, in Sect. 4.5, we analyze the problems related to feedback.

4.1 Analyzing the Early Stages of Idea Propagation

New good ideas sometimes propagate too slowly. To speed up their propagation, we need to have a quantitative understanding of how ideas propagate. An intuitive understanding of ideas propagation has led to several reasonable first-approximation mathematical models. These models provide a good description of idea propagation on the later stages, when the ideas have already been adopted by a reasonably large number of people. However, at the critically important early stages, these models are not perfect: these models predict a linear growth with time, while empirical growth data is often better described by a power law.

In the first subsection of this section, we provide an intuitive theoretical explanation of the observed power-law growth. In the second subsection, we describe our case study that confirms the ubiquity of the power law.

L.O. Lerma and V. Kreinovich, *Towards Analytical Techniques for Optimizing Knowledge Acquisition, Processing, Propagation, and Use in Cyberinfrastructure and Big Data*, Studies in Big Data 29, DOI 10.1007/978-3-319-61349-9_4

4.1.1 Power Law Model of Knowledge Propagation: Theoretical Explanation

In this section, we provide a theoretical explanation for the power law model of knowledge propagation. The results from this section were first published in [1].

Propagation of new tools and new ideas—one of the main ways science and technologies progress. Science and technology are progressing at an enormous speed. New ideas appear all the time, new tools are being designed all the time that enable us to do things that we could not do before—and do them faster, more reliably, and more efficiently.

It is extremely important to come up with new ideas, to design new tools, but mere design is not enough: it is important to make sure that these ideas and tools do not stay with their inventors, that they are widely adopted and thus propagate.

Current first approximation model of ideas propagation. We would like to know how ideas propagate, i.e., how the number $n(t)$ of people who use the new idea grows with time t.

The main current model of idea propagation (see, e.g., [2–12] and references therein) is as follows. For the idea to spread, people who have not yet adopted the idea must learn about it—either from the original announcement or from people who already use this idea. The probability that a new person will learn this new idea can thus be estimated as $a + b \cdot n(t)$, where a is the probability to learn this idea from the original announcement, and b is the probability to encounter one of the followers. Out of the total population of N people, $N - n(t)$ not-yet-users are exposed to this learning. Since the probability of each of them learning about the new idea is proportional to $a + b \cdot n(t)$, the total number of people who learn about the new idea is proportional to

$$(a + b \cdot n(t)) \cdot (N - n(t)).$$

Thus, we arrive at the differential equation

$$\frac{dn}{dt} = c \cdot (a + b \cdot n(t)) \cdot (N - n(t)),$$

where c is the corresponding proportionality coefficient. Thus, we get

$$\frac{dn}{dt} = (A + B \cdot n(t)) \cdot (N - n(t)), \tag{4.1.1}$$

where $A \stackrel{\text{def}}{=} c \cdot a$ and $B \stackrel{\text{def}}{=} c \cdot b$.

Many refinements of this model have been proposed (see, e.g., [8]), but the model (4.1.1) remains the main first approximation model of knowledge propagation.

Solution to the first approximation model. By moving all the terms containing n to the left-hand side and all the terms containing time t to the right-hand side, we conclude that

$$\frac{dn}{(A + B \cdot n) \cdot (N - n)} = dt. \tag{4.1.2}$$

Here, as one can easily check,

$$\frac{1}{(A + B \cdot n) \cdot (N - n)} = \frac{1}{A + B \cdot N} \cdot \left(\frac{B}{A + B \cdot n} + \frac{1}{N - n} \right). \tag{4.1.3}$$

Thus, the left-hand side of the formula (4.1.2) takes the form

$$\frac{1}{A + B \cdot N} \cdot \left(\frac{B \cdot dn}{A + B \cdot n} + \frac{dn}{N - n} \right) =$$

$$\frac{1}{A + B \cdot N} \cdot \left(\frac{d(A + B \cdot n)}{A + B \cdot n} - \frac{d(N - n)}{N - n} \right). \tag{4.1.4}$$

So, the integral of this left-hand side takes the form

$$\frac{1}{A + B \cdot N} \cdot (\ln(A + B \cdot n) - \ln(N - n)) = \frac{1}{A + B \cdot N} \cdot \ln \left(\frac{A + B \cdot n}{N - n} \right). \tag{4.1.5}$$

Hence, integrating both sides of the Eq. (4.1.2), we get

$$\frac{1}{A + B \cdot N} \cdot \ln \left(\frac{A + B \cdot n}{N - n} \right) = t + C, \tag{4.1.6}$$

where C is the integration coefficient. Therefore,

$$\ln \left(\frac{A + B \cdot n}{N - n} \right) = k \cdot t + c, \tag{4.1.7}$$

where $k \stackrel{\text{def}}{=} A + B \cdot N$ and $c \stackrel{\text{def}}{=} k \cdot C$. Raising e to the (equal) left- and right-hand sides of the Eq. (4.1.7), we get

$$\frac{A + B \cdot n}{N - n} = C' \cdot \exp(k \cdot t), \tag{4.1.8}$$

where $C' \stackrel{\text{def}}{=} \exp(c)$. Multiplying both sides of this equation by $N - n$, we get

$$A + B \cdot n = N \cdot C' \cdot \exp(k \cdot t) - n \cdot C' \cdot \exp(k \cdot t), \tag{4.1.9}$$

hence

$$n \cdot (B + C' \cdot \exp(k \cdot t)) = C' \cdot \exp(k \cdot t) - A, \tag{4.1.10}$$

and, thus,

$$n(t) = \frac{C' \cdot \exp(k \cdot t) - A}{B + C' \cdot \exp(k \cdot t)}. \quad (4.1.11)$$

If we start measuring time from the moment the idea was launched, so that $n(0) = 0$, then we conclude that $C' = A$ and thus, the formula (4.1.11) takes the form

$$n(t) = \frac{A \cdot (\exp(k \cdot t) - 1)}{B + A \cdot \exp(k \cdot t)}. \quad (4.1.12)$$

Initial propagation of a new idea: asymptotic description. Once the idea has spread, it will continue spreading; the most critical period is right after the idea's appearance, when $t \approx 0$. For such t, asymptotically, the first approximation model (4.1.12) implies that

$$n(t) \approx c \cdot t, \quad (4.1.13)$$

where $c \stackrel{\text{def}}{=} \frac{A \cdot k}{B + A}$ i.e., that $n(t)$ linearly grows with time t.

Empirical data seems to be inconsistent with this asymptotics. While for medium and large times t, the first approximation model (4.1.12) is in a reasonably good accordance with data, for small t, the empirical data shows a clearly non-linear behavior (see, e.g., [3, 13]), a behavior which is better described by a power function

$$n(t) \approx c \cdot t^a \quad (4.1.14)$$

for some $a \neq 1$.

What we do in this section. In this section, we provide a simple intuitive model which explains such power-law growth.

Main idea behind our explanation: a qualitative description. To describe how a new idea propagates, let us consider one specific tool (or idea) aimed at solving problems from a specific class. For example, this tool may be a new (e.g., more efficient) software for solving large systems of linear equations.

Any person who sometimes solves the problem from the given class is a potential user of this tool. We start with the initial situation, in which only the author of the tool knows it and uses it. Eventually, other potential users start learning and using this tool.

When does a potential user start learning the new tool? On the one hand, there are clear benefits in learning a new tool: once a person learns the new tool, he or she can solve the problems from the corresponding class more efficiently.

- For example, efficiency may mean faster computations. In this case, the user will be able to solve large systems of linear equations faster. This will save the time needed to solve such systems, and enable the user to get the results faster.
- Alternatively, efficiency may mean that the user may be able to use fewer processors of a multi-processor computer system to solve the same problem—so, if the user

pays for the computer time, he or she will be able to save some money by using this new software.

On the other hand, new tools, new ideas are not always easy to learn. One needs to invest some effort—e.g., time—into learning the new tool. A potential user will start learning the new tool only if the expected benefits exceed the investment needed to learn this tool. So, to figure out when a particular user will start learning the tool, we need to be able to estimate both potential benefits and the required investment.

The potential benefits of using a tool depend on how often it will be used.

- Some potential users solve the corresponding problems very frequently. For such users, a potential benefit may be large.
- Other users encounter the corresponding problems rarely. For such users, the potential benefit of learning the new tool may be small.

In our analysis, we need to take this difference into account.

To estimate the difficulty of learning the tool, we need to take into account that this difficulty depends on how many people have already learned it. If a tool is currently used only by a few folks, it is more difficult to learn it: if there is a question about this tool, it is not so easy to find someone who knows the answer. On the other hand, if the tool is widely used, learning this tool is much easier: when there is a question, one of the nearby colleagues who is already using this tool can answer.

For example, in a Computer Science department, it is easy for someone to learn one of the widely used languages such as C++ or Java: whatever question may arise, there are plenty of people around who know these languages already. On the other hand, a new operating system—e.g., a new version of Windows—may be simpler to use than C++, but in the beginning, it is not so easy to learn—since in the beginning, very few people have an experience of using it and therefore, it is difficult to find help if a problem arises.

Let us show how these qualitative ideas can be translated into a quantitative model.

Heavy users versus light users. As we have mentioned, a user will start learning the new tool only if the expected benefit of its use exceeds the expenses needed to learn this tool. For each user, the expected benefit b of using the tool is proportional to the number x of the corresponding problems (per unit time) that this user encounters: $b = C \cdot x$, for some proportionality constant C (that describes the benefit of using the tool to solve a single problem). From this viewpoint, each user can be characterized by the corresponding value x.

Let $L(t)$ describe the cost of learning the tool at moment t. In this notation, at each moment of time t, a potential user—characterized by the value x—will start learning the tool if the benefit $C \cdot x$ exceeds the cost $L(t)$: $C \cdot x > L(t)$. This condition can be equivalently described as $x > x_0(t)$, where we denoted $x_0(t) \stackrel{\text{def}}{=} \frac{L(t)}{C}$. This ratio $x_0(t)$ serves as a threshold:

- "heavy users", i.e., users for which $x > x_0(t)$, will start adopting the tool, while
- "light users", i.e., users for which $x < x_0(t)$, will continue using previous tools.

Thus, at each moment of time, the state of propagation can be characterized by a single value—this threshold value $x_0(t)$.

Distribution of users. To describe how knowledge propagates, we need to know how many users are there with different levels of usage x. In many practical problems, the distribution is described by the power law (see, e.g., [14, 15]), in which the proportion $P(x \geq X)$ of objects x for which x exceeds a given threshold X is determined by a formula

$$P(x \geq X) = C_0 \cdot X^{-\alpha} \tag{4.1.15}$$

for some constants C_0 and α. The ubiquitous character of power laws was popularized by Benoit Mandelbrot in his fractal theory; see, e.g., [16].

How easier is to to learn a new tool when we already have a given number n of users. In the beginning, learning a new tool is not very easy, but ultimately, tools and techniques become relatively easy to learn. For example:

- calculus used to be a great 17 century achievement, accessible only to a few great minds;
- however, nowadays, many kids study elements of calculus already in high school.

The reason why, in the beginning, learning a new tool is not easy is that a person learning the new tool *can* go astray (and *goes* astray). The more advice we get, the more accurately we understand what needs to be done—i.e., crudely speaking, the more accurate is the direction in which we are going—and thus, the smaller amount of effort will be wasted. The resulting amount of effort can be viewed as proportional to the inaccuracy with which we know the direction in which to go in learning the tool.

To find this direction, we can use the advice and expertise of the existing users. If we have n users that we can consult, this means that we have n estimates for the desired direction. In general, according to statistics (see, e.g., [17]), if we have n similar independent estimates of the same quantity, then, by taking their average, we can get a combined estimate which is $\sqrt{n}$ times more accurate than each of the individual estimates. Thus, it is reasonable to assume that when we have n users, the amount of effort needed to learn the tool is (approximately) equal to $\frac{b}{\sqrt{n}}$ for some constant b.

Resulting dynamics of propagation. As we have mentioned, a person starts learning a new tool if the expected benefit of its use exceeds the cost of learning. Once we have n users, the cost of learning is equal to $\frac{b}{\sqrt{n}}$. The expected benefit of leaning the tool is proportional to the average number of problems encountered by the potential user, i.e., to the number x; in other words, this benefit can be described as $a \cdot x$ for some constant a. So, at this stage, only persons for which $a \cdot x \geq \frac{b}{\sqrt{n}}$ have an incentive to study this tool. This condition can be described equivalently as $x \geq X_0$, where we denoted $X_0 \stackrel{\text{def}}{=} \dfrac{b}{a \cdot \sqrt{n}}$. According to the power-law distribution, out of N who may be potentially interested in this tool, the total number of persons who have an incentive to study this tool is equal to

$$N \cdot P(X > X_0) = N \cdot C_0 \cdot X_0^{-\alpha} = N \cdot C_0 \cdot \left(\frac{b}{a \cdot \sqrt{n}}\right)^{-\alpha} = c_1 \cdot n^{\alpha/2}, \quad (4.1.16)$$

for an appropriate constant $c_1 \stackrel{\text{def}}{=} N \cdot C_0 \cdot \left(\dfrac{b}{a}\right)^{-\alpha}$.

The rate $\dfrac{dn}{dt}$ with which the number of users n increases is proportional to the number of potential users who study the new tool, i.e., to the number of persons who have an incentive to study this tool. Thus, we conclude that

$$\frac{dn}{dt} = c_2 \cdot n^{\alpha/2} \quad (4.1.17)$$

for some constant c_2.

Moving terms containing n to the left-hand side and terms containing t to the right-hand side, we get

$$n^{-\alpha/2} \cdot dn = c_2 \cdot dt. \quad (4.1.18)$$

Now, we can integrate both sides. The result of this integration depends on the value $\alpha/2$.

When $\alpha/2 = 1$, integration leads to

$$\ln(n) = c_2 \cdot t + C, \quad (4.1.19)$$

where C is the integration constant. We want to describe the starting period of idea propagation, when $n(t) = 0$ for $t = 0$. For $n = 0$, however, the left-hand side of (4.1.19) is infinite, while the right-hand side is finite. Thus, we cannot have $\alpha/2 = 1$.

When $\alpha/2 \neq 1$, integration of (4.1.18) leads to

$$\frac{1}{1-\alpha/2} \cdot n^{1-\alpha/2} = c_2 \cdot t + C. \quad (4.1.20)$$

We want to satisfy the requirement that $n(t) = 0$ when $t = 0$. When $t = 0$, the right-hand side of the formula (4.1.20) is equal to C. When $n = 0$, the value $n^{1-\alpha/2}$ is equal to 0 when $1 - \alpha/2 > 0$ and to ∞ when $1 - \alpha/2 < 0$. Since $C < \infty$, the condition that $n(t) = 0$ when $t = 0$ can only be satisfied when $1 - \alpha/2 > 0$. In this case, for $t = 0$, the formula (4.1.19) takes the form $0 = C$. Substituting $C = 0$ into the formula (4.1.19), we conclude that

$$\frac{1}{1-\alpha/2} \cdot n^{1-\alpha/2} = c_2 \cdot t, \quad (4.1.21)$$

hence

$$n^{1-\alpha/2} = (c_2 \cdot (1 - \alpha/2)) \cdot t. \tag{4.1.22}$$

Raising both side by the power $a \stackrel{\text{def}}{=} 1/(1 - \alpha/2)$, we conclude that

$$n(t) = c \cdot t^a, \tag{4.1.23}$$

where $c \stackrel{\text{def}}{=} (c_2 \cdot (1 - \alpha/2))^a$. This is exactly the formula that we wanted to explain.

Summary of the results. So, the above light user–heavy users model indeed explains the observed power-law growth of the number of adoptees of a new idea.

4.1.2 Testing a Power Law Model of Knowledge Propagation: Case Study of the Out of Eden Walk Project

To improve teaching and learning, it is important to understand how knowledge propagates. In general, when a new piece of knowledge is introduced, people start learning about it. Since the potential audience is limited, after some time, the number of new learners starts to decrease. Traditional models of knowledge propagation are based on differential equations; in these models, the number of new learners decreases exponentially with time. As we have mentioned, recently, a new power law model for knowledge propagation was proposed. In this model, the number of learners decreases much slower, as a negative power of time. In this paper, we compare the two models on the example of readers' comments on the Out of Eden Walk, a journalistic and educational project in which informative messages ("dispatches") from different parts of the world are regularly posted on the web. Readers who learned the new interesting information from these dispatches are encouraged to post comments. Usually, a certain proportion of readers post comments, so the number of comments posted at different times can be viewed as a measure characterizing the number of new learners. So, we check whether the number of comments is consistent with the power law or with the exponential law. To make a statistically reliable conclusion on which model is more adequate, we need to have a sufficient number of comments. It turns out that for the vast majority of dispatches with sufficiently many comments, the observed decrease is consistent with the power law (and none of them is consistent with the exponential law).

Results from this section first appeared in [18, 19].

What we do in this section. In this section, we empirically check whether power law is indeed a better description of knowledge propagation than models based on differential equations. For that purpose, we consider an example—comments on the messages ("dispatches") posted as a part of the Out of Eden Walk project [20]; see also [21–23].

Out of Eden Walk project: a description. Commenced on January 10th, 2013 in Ethiopia, the Out of Eden Walk is a 7-year, 21,000 mile long, storytelling journey created by two-time Pulitzer Prize winning journalist Paul Salopek. This project is sponsored by the National Geographic Society. Reports from this journey regularly appear in the National Geographic magazine, in leading newspapers such as New York Times, Washington Post, Chicago Tribune, Los Angeles Times, and on the US National Public Radio (NPR).

This project has important educational and knowledge propagation goals. The Out of Eden Walk is a very ambitious project, its main objective is to enhance education and knowledge propagation as main features of journalism, to reinvent digital reporting in the age of nano-headlines by embracing the concept of *slow journalism*: revealing human stories and world events from the ground, at a walking pace. The slow journalism of the Out of Eden Walk is immersive and sustained reporting, yet conveyed through the state-of-the-art digital platforms, with presence on the web, on Facebook, on Twitter, and in the traditional media.

The project has largely succeeded in these goals: now in its third year, the website has thousands of followers worldwide, not counting Facebook, Twitter, and other followers. Over 200 schools worldwide regularly use Salopek's reports as an education tool, to enrich the students' understanding of different worldwide cultures.

Out of Eden Walk project: technical details. After visiting a new geographic area, Paul Salopek selects an important topic related to this area and publishes a *dispatch* describing his impressions, experiences, and thoughts. As of January 2015—the time of our analysis—there were close to 100 dispatches.

Followers are welcome to add comments after each dispatch. After two weeks, each dispatch gathers from 15 to more than 250 comments. Many of these comments are made by teachers and students who use these dispatches as part of their learning experience.

Sometimes, Paul Salopek replies to some of these comments, often by providing additional details about the story. These replies, in their turn, elicit more comments, etc. All these comments are part of the knowledge propagation process.

What we do. In our analysis, we traced, for each dispatch, how the number of comments made by the readers changes with time. This number reflects how the knowledge contained in a dispatch propagates with time.

Specifically, as we mentioned earlier, we check whether this propagation is better described by a traditional model based on differential equations or by a power law. To make this comparison, let us first recall the formulas describing these two approaches to quantifying knowledge propagation.

The power law model. The power law formulas predict that the number of comments $r(t)$ decreases with t as $r(t) = A \cdot t^{-\alpha}$. This model has two parameters A and $\alpha > 0$.

In practice, after a large period of time t, the number of new comments decreases to practically 0—and this is exactly what the power law predicts, since $A \cdot t^{-\alpha} \to 0$ as $t \to \infty$.

The traditional model. The traditional description of knowledge propagation is based on first order differential equations. In particular, a general way to describe how the number of comments $r(t)$ changes with time is to use a differential equation $\frac{dr}{dt} = -f(r)$, for an appropriate function $f(r)$.

As we have mentioned, in practice, after a large period of time t, the number of new comments decreases to practically 0. In this case, we have $r(t) \approx 0$ and $\frac{dr}{dt} \approx 0$. So, we should have $f(0) = 0$.

In principle, we can have models of different complexity. We can have models with a linear function $f(r)$, we can have models with a more general quadratic dependence $f(r)$, etc. To make a fair comparison, we should select a class of models which is characterized by the same number of parameters as the power law model—otherwise, the traditional model will be more accurate just because we allow it to use more parameters to adjust to the data. Let us start with the simplest case of a linear function $f(r)$. A general linear function with $f(0) = 0$ has the form $f(r) = \alpha \cdot r$ for some α. For this function $f(r)$, the above differential equation has a 2-parametric family of solutions $r(t) = A \cdot \exp(-\alpha \cdot t)$.

For quadratic functions $f(r) = \alpha \cdot r + c \cdot r^2$, we already have a 3-parametric family of solutions. So, to keep our comparison fair, in this section, we use the exponential model $r(t) = A \cdot \exp(-\alpha \cdot t)$ as a traditional model of knowledge propagation.

How we compare the two models: selecting a time period. Our observation is that once a dispatch is posted, there is a short period with practically no comments, then the bulk of the comments start, first with a big burst in comments, and then usually gradually decreasing.

Both power law and exponential law describe only how the number of comments decreases with time. Thus, to compare the observations with the model, we start with the day on which the most comments were posted, and considers this day and several days after that. In most cases, the vast majority of comments are posted within the first month, so we limited our data to 30 consecutive days (starting with the day in which the largest number of comments were posted).

Finding parameters of the model: Least Squares approach. In our analysis, for each of the two models, we first find the values of the parameters leading to the best fit, and then check how good is the resulting fit.

Let us start with the power law model. The number of responses fluctuates, so clearly responding is a random process. Let $p(t)$ denote the probability with which a person responds at moment t. The overall numbers of responses $r(t)$ can be viewed as a the sum $r(t)_1 + \ldots + r(t)_N$, where $r(t)_i$ is the number of responses coming from the i-th reader at moment t. Since the probability of a person responding is equal to $p(t)$, this means that each value r_i is equal to 1 with probability $p(t)$ and to 0 with the remaining probability $1 - p(t)$. One can easily see that the expected value $E[r_i]$ is then equal to $p(t)$, and the variance $V[r_i]$ is equal to $p(t) \cdot (1 - p(t))$. Since we assume that all the respondents are independent, the expected value of the sum is equal to the sum of the expected values, and the variance of the sum is equal to the sum of variances. Thus, the expected value is equal to $N \cdot p(t)$, and the variance is equal to $N \cdot p(t) \cdot (1 - p(t))$.

It is reasonable to estimate the probability $p(t)$ by the corresponding frequency, i.e., as the ratio $\frac{r(t)}{N}$. In this case, the variance can be estimated as

$$V(t) = N \cdot \frac{r(t)}{N} \cdot \left(1 - \frac{r(t)}{N}\right).$$

Here, $r(t) \ll N$, so $1 - \frac{r(t)}{N} \approx 1$, and $V(t) \approx N \cdot \frac{r(t)}{N} = r(t)$. Since the standard deviation is equal to the square root of the variance, we conclude that

$$r(t) \approx r_0(t), \text{ with accuracy } \sigma(t) = \sqrt{r(t)},$$

where $r_0(t)$ denotes the corresponding model.

Finding parameters of the model: case of power model. For the power model $r_0(t) = A \cdot t^{-\alpha}$, the above formula takes the form

$$r(t) \approx A \cdot t^{-\alpha}, \text{ with accuracy } \sigma(t) = \sqrt{r(t)}. \quad (4.1.24)$$

For large N, due to the Central Limit theorem, the distribution of an approximation error is close to Gaussian, so, in principle, we can find the parameters A and α by using the Maximum Likelihood method, which in this case takes the form

$$\sum_t \frac{(r(t) - A \cdot t^{-\alpha})^2}{\sigma^2(t)} = \sum_t \frac{(r(t) - A \cdot t^{-\alpha})^2}{r(t)} \to \min .$$

From the computational viewpoint, however, this approach has a limitation. Such a Gaussian-based Maximum Likelihood (= Least Squares) approach is usually applied to situations when the model linearly depends on the parameters. In our case, however, the objective function is a quadratic function of these parameters, and so, by differentiating this objective function with respect to all these parameters and equating the resulting partial derivatives to 0, we get an easy-to-solve system of linear equations. In our case, the dependence of the model on the parameters A and α is strongly non-linear, and, as a result, we end up with a mode-difficult-to-solve system of nonlinear equations. To simplify computations, it is therefore desirable to reduce our problem to the case when the model linearly depends on parameters.

It is known that in log-log scale, the power law becomes a linear dependence. Specifically, if we take logarithms of both sides of the formula $r(t) = A \cdot t^{-\alpha}$, then we get $\ln r(t) = \ln(A) - \alpha \cdot \ln(t)$.

To use this fact, we need to find out how the inaccuracy in $r(t)$ is transformed into the inaccuracy with which we know $\ln(r(t))$. We know that the inaccuracy $\Delta r(t)$ has standard deviation $\sqrt{r(t)}$. When this inaccuracy is small, for $\ln(r(t))$, we have $\frac{\Delta(\ln(r(t)))}{\Delta r(t)} \approx \frac{d(\ln(r(t))}{dr(t)} = \frac{1}{r(t)}$. Thus, in this approximation, $\Delta(\ln(r(t)) =$

$\frac{\Delta r(t)}{r(t)}$. In general, when we multiply a random variable ξ by a positive constant c, its standard deviation $\sigma[\xi]$ is multiplied by the same constant: $\sigma[c \cdot \xi] = c \cdot \sigma[\xi]$. Thus, we conclude that $\sigma[\ln(r(t))] = \frac{\sigma[r(t)]}{r(t)}$. Since we already know that $\sigma[r(t)] = \sqrt{r(t)}$, we thus conclude that $\sigma[\ln(r(t))] = \frac{1}{\sqrt{r(t)}}$. For the power model, we have $\ln(r(t)) \approx \ln(A) - \alpha \cdot \ln(t)$, so we have

$$\ln(r(t)) \approx \ln(A) - \alpha \cdot \ln(t) \text{ with accuracy } \sigma = \frac{1}{\sqrt{r(t)}}.$$

If we multiply both sides of this approximate equality by $\sqrt{r(t)}$, we conclude that

$$\sqrt{r(t)} \cdot \ln(r(t)) \approx a \cdot \sqrt{r(t)} + b \cdot \ln(t) \cdot \sqrt{r(t)} \text{ with accuracy } \sigma = 1,$$

where we denoted $a \stackrel{\text{def}}{=} \ln(A)$ and $b \stackrel{\text{def}}{=} -\alpha$. For this problem, the Maximum Likelihood (= Least Squares) method means minimizing the sum

$$\sum_t \left(\sqrt{r(t)} \cdot \ln(r(t)) - a \cdot \sqrt{r(t)} - b \cdot \ln(t) \cdot \sqrt{r(t)}\right)^2.$$

Once we find the corresponding values of a and b, we can then find $A = \exp(a)$ and $\alpha = -b$.

Finding parameters of the model: case of exponential model. For the exponential model $r_0(t) = A \cdot \exp(-\alpha \cdot t)$, the formula for finding the parameters takes the form

$$r(t) \approx A \cdot \exp(-\alpha \cdot t), \text{ with accuracy } \sigma(t) = \sqrt{r(t)}. \tag{4.1.25}$$

Similarly to the case of the power model, we can conclude that the distribution of an approximation error is close to Gaussian, so, we can find the parameters A and α by using the Maximum Likelihood method, which in this case takes the form

$$\sum_t \frac{(r(t) - A \cdot \exp(-\alpha \cdot t))^2}{\sigma^2(t)} = \sum_t \frac{(r(t) - A \cdot \exp(-\alpha \cdot t))^2}{r(t)} \to \min.$$

The exponential law $r(t) = A \cdot \exp(-\alpha \cdot t)$ becomes linear if we consider the dependence of $\ln(r(t))$ on time t: $\ln(r(t)) = \ln(A) - \alpha \cdot t$.

We already know, from our analysis of the power model, that $\sigma[\ln(r(t))] = \frac{\sigma[r(t)]}{r(t)}$. Thus, we have

$$\ln(r(t)) \approx \ln(A) - \alpha \cdot t \text{ with accuracy } \sigma = \frac{1}{\sqrt{r(t)}}.$$

If we multiply both sides of this approximate equality by $\sqrt{r(t)}$, we conclude that

$$\sqrt{r(t)} \cdot \ln(r(t)) \approx a \cdot \sqrt{r(t)} + b \cdot t \cdot \sqrt{r(t)} \text{ with accuracy } \sigma = 1,$$

where we denoted $a \stackrel{\text{def}}{=} \ln(A)$ and $b \stackrel{\text{def}}{=} -\alpha$. For this problem, the Maximum Likelihood (= Least Squares) method means minimizing the sum

$$\sum_t \left(\sqrt{r(t)} \cdot \ln(r(t)) - a \cdot \sqrt{r(t)} - b \cdot t \cdot \sqrt{r(t)}\right)^2.$$

Once we find the corresponding values of a and b, we can then find $A = \exp(a)$ and $\alpha = -b$.

From the traditional Least Squares to the robust ℓ^1 method for parameter estimation. When a model predicts the values E_i and observations are O_i, the traditional Least Squares estimate selects the parameters for which the sum of the squares of the differences $\sum_i (O_i - E_i)^2$ is the smallest possible. As we have mentioned, this works well if the difference $O_i - E_i$ is normally distributed.

In practice, in addition to a normally distributed differences $O_i - E_i$, we also have outliers. For example, a measuring instrument may malfunction, generating a number which is far away from the actual value of the measured quantity. Such outliers can drastically change the Least Squares estimate.

For example, if the model is a constant $E_i = \text{const}$, the Least Square estimate for this constant is simply the arithmetic mean of all observed values $\frac{1}{n} \cdot \sum_{i=1}^{n} O_i$. If there are no outliers, this works well. For example, if the actual value is 0, and standard deviation is $\sigma = 1$, then after 100 observations, we get 0 with an accuracy of $\frac{\sigma}{\sqrt{n}} \approx 0.1$. However, if due to a malfunction, one of the recordings is 10000, we get the average 100.

In situations when outliers are possible, it is therefore reasonable to use methods which are less sensitive to outliers. Such methods are known as *robust*; see, e.g., [24]. One of the most widely used robust methods is the ℓ^1 *-method*, when we select parameters for which the sum of absolute values $\sum_i |O_i - E_i|$ is the smallest possible.

For example, in the above example when the model is a constant, the ℓ^1 methods results in selecting a median of all the observation instead of the arithmetic mean, and one can easily check that the median is much less sensitive to outliers.

To take into account the possibility of outliers, in this section, in addition to using the Least Squares method to find the parameters of the model, we also use an ℓ^1 method.

How to check whether the data fits a model. Once we have found the parameters of a model, we can apply the Pearson's chi-squared test to check whether the data fits the corresponding formula. This test checks whether the number of events $O_1, \ldots, O_n$

that occurred in n different situations is consistent with the model that predicts that, one average, E_i events will occur in situation i. To apply this test, we compute the sum $\mu = \sum_{i=1}^{n} \frac{(O_i - E_i)^2}{E_i}$. This distribution of this value is close to the chi-squared distribution with $n - n_{\text{param}}$ parameters, where n_{param} is the number of parameters that was determined based on this distribution (in our case, $n_{\text{param}} = 2$). Then, we compare a p-value by comparing the value of the statistic to a chi-squared distribution, and use this p-value to decide whether the observations fit the model.

In our case, different situations i correspond to different moments of time t, the expected values come from the corresponding model $E_i = r_0(t)$, and the observed values are $O_i = r(t)$. Thus, for this situation, Pearson's chi-squared test means computing the expression

$$\chi^2 \stackrel{\text{def}}{=} \sum_t \frac{(r(t) - r_0(t))^2}{r_0(t)}.$$

In particular, for the power model, we compute the value

$$\chi_p^2 \stackrel{\text{def}}{=} \sum_t \frac{(r(t) - A \cdot t^{-\alpha})^2}{A \cdot t^{-\alpha}}.$$

For exponential model, we compute the value

$$\chi_e^2 \stackrel{\text{def}}{=} \sum_t \frac{(r(t) - A \cdot \exp(-\alpha \cdot t))^2}{A \cdot \exp(-\alpha \cdot t)}.$$

Based on this value of χ^2 and on the number $n-2$ of degrees of freedom, we compute the corresponding p-value p_p or p_e.

The values of χ^2 and p based on ℓ^1-estimates will be denoted, correspondingly, by $\chi_{p,1}^2$, $\chi_{e,1}^2$, $p_{p,1}$, and $p_{e,1}$.

If this p-value is smaller than some threshold p_0 (usually, $p_0 = 0.05$), then we can conclude that the data is inconsistent with the corresponding model, and the model is rejected. Otherwise, if the p-value is greater than or equal to the threshold p_0, we conclude that the data is consistent with the model.

Selecting dispatches. When the sample size is small, both models fit. Empirically, we have found that the models can be separated if we have at least 50 comments. Some dispatched are shorter than others; these dispatches are marked as "trail notes". None of these trail notes has 50 or more comments; so, we only considered "proper" dispatches, i.e., dispatches which are not trail notes.

Our interest is in analyzing knowledge propagation. New dispatches appear all the time, and new comments are added all the time. Thus, the more recent the dispatch, the more probable it is that new comments will be added and therefore, that the available comments do not yet present a final description of how the corresponding knowledge propagates. So, in our analysis, we concentrated on the earliest dispatches, for which the picture of knowledge propagation is (most probably) complete.

Specifically, to compare the two models, out of the 25 earlier "proper" dispatches, we selected all the dispatches which by February 1, 2015, had at least 50 comments. There were seven such dispatches; these dispatches are listed in the following table.

For each dispatch, there is usually a few days delay until the bulk of the comments appear. At first, there is a burst of comments, then the number of comments gradually decreases, and after a month, very few new comments appear. So, for our analysis, we limited our data to a time period that:

- starts on the first day when the largest number of comments appear, and
- ends 30 days after the posting of the original dispatch.

Comparison results. The result of analyzing the selected ten dispatches are given in the following table; here, N_c is the number of comments, and all the p-values are rounded to two digits. P-values exceeding 0.05—that indicate that the model is consistent with the data—are underlined.

The ℓ^1 values corresponding to the exponential model are not given, but the resulting p-values are similar to the values corresponding to the Least Squares estimates.

Dispatch Title	N_c	χ_p^2	$\chi_{p,1}^2$	χ_e^2	p_p	$p_{p,1}$	p_e
Let's Walk	271	30.6	30.0	31,360	<u>0.33</u>	<u>0.37</u>	0.00
Sole Brothers	61	22.1	22.8	83	<u>0.76</u>	<u>0.74</u>	0.00
The Glorious Boneyard	59	16.3	18.6	262	<u>0.96</u>	<u>0.91</u>	0.00
The Self-Love Boat	67	63.1	60.0	124	0.00	0.00	0.00
Go Slowly–Work Slowly	91	33.0	31.5	821	<u>0.24</u>	<u>0.29</u>	0.00
The Camel and the Gyrocopter	52	28.4	24.6	72	<u>0.45</u>	<u>0.65</u>	0.00
Lines in Sand	69	21.4	18.3	89	<u>0.81</u>	<u>0.92</u>	0.00

Conclusions:

- None of the dispatches is consistent with the exponential law.
- For both the Least Squares method and a more robust ℓ^1-method to estimate the parameters, the vast majority of the dispatches is consistent with the power law.

So, this data supports the power law model in comparison with the more traditional exponential model.

Discussion. A possible reason why comments on some dispatches fit neither the power law nor the exponential law is that Salopek sometimes replies to the comments, and these replies trigger another wave of comments. As a result, some observed distributions of comments over time are bimodal or close to bimodal.

Summary. To improve teaching and learning, it is important to know how knowledge propagates. Traditional models of knowledge propagation are similar to differential-equations-based models of propagation in physics. Recently, an alternative fractal-motivated power-law model of knowledge propagation was proposed, that, in several cases, provides a more adequate description of knowledge propagation. In this section, we compare this model with the traditional model on the example of the comments to the Out of Eden Walk project, an ambitious journalistic and educational project aimed at educating the general audience about different societies around the world, their culture, their history, etc.

It turns out that for the related data, the power law is indeed a more adequate description:

- for the vast majority of dispatches, the dependence of number of comments on time is consistent with the power law, while
- the differential equations-motivated exponential law is not consistent with any of this data sets.

This shows that the fractal-motivated power law is indeed a more adequate description of knowledge propagation—to be more precise, a more adequate first approximation to describing knowledge propagation.

Possible future work. There are many possibilities to expand our analysis:

- we can further analyze our power law models,
- we can go beyond the power law models and try to get models which are even more adequate for describing knowledge propagation,
- instead of simply counting comments corresponding to different dispatches, we can also look into the substance of these comments and dispatches, and finally,
- we can use information beyond web-posted comments.

Let us describe these possibilities one by one.

Possible further analysis of the power law models corresponding to different dispatches. First, it would be interesting to *further analyze* our power law results. In our analysis, we simply checked whether the dependence is described by the power law or not. However, the power laws come with different parameters α. The larger α, the faster the number of comments decreases with time. It would be interesting to check whether the values α corresponding to different dispatches depend on the overall number of comments. Based on the few dispatches that we analyzed, it appears that when we have more initial comments, then the decrease is slower, but so far, we do not have enough data to rigorously confirm this observation.

Beyond power law models. As we have mentioned earlier, the power law explains the time distribution of comments only for a little more than a half of dispatches. It is therefore desirable to design models that would explain the time dependence for all the dispatches—or at least for a large portion of dispatches. There are two possible ways to get a better fit.

One possibility is to consider *more complex* models. A natural way to do it is to take into account that a power law is linear in log-log scale, it leads to a linear dependence of $\ln(r(t))$ on $\ln(t)$:

$$\ln(r(t)) \approx \ln(A) - \alpha \cdot \ln(t);$$

thus, a natural idea is to see if a quadratic dependence

$$\ln(r(t)) \approx \ln(A) - \alpha \cdot \ln(t) + \beta \cdot (\ln(t))^2$$

leads to a better fit.

Another possibility is to take into account some *external events* that affect the number of comments: for example, the number of comments spikes when Salopek replies to comments, when he appears on National Public Radio, or when an article of his is published in the *National Geographic* magazine.

Probably, both ideas need to be implemented to get a better fit with the observed number of comments to different dispatches.

Beyond counting the number of comments. In our analysis, we simply counted the number of comments. It is desirable to also take into account the substance of these comments—and also the substance of the dispatch.

Our main interest is in the educational applications. From the educational viewpoints, *not all comments are equal*:

- some comments simply praise the dispatch, without any explicit indication that the replier learned something new from it;
- other comments explicitly indicate that the replier learned some new information from the dispatch;
- a few comments go even further, indicating that the replier plans to inform their friends and colleagues about some material from this dispatch—in particular, some repliers who are teachers plan to use this material in their classes.

It would be interesting to analyze how the numbers of such "learning" and "teaching" comments changes with time.

Similarly, *not all dispatches are equal.* Some topics—e.g., the dispatches about the beasts of burden such as camels or mules—caused many comments, and these comments come a few days after the dispatch is posted. On the other hand, other topics cause fewer replies. It would be interesting to analyze which types of dispatches elicit more comments. This may be useful in teaching, since eliciting comments from students is a known way to improve their learning.

It is also desirable to use *information about the repliers*:

- The majority of repliers sign their comments with their *names*. It would be interesting to trace how many repliers for each dispatch are new and how many also replied to one or more of the previous dispatches; this will show how the audience changes in time.
- Some repliers indicate their *geographic location*. It would be interesting to analyze the geographic distribution of comments, and to see how the geographic distribution of comments to a given dispatch depends on the geographic area that is the subject of this dispatch.

Beyond web-based replies. Finally, we should take into consideration that, in addition to comments posted on the website, repliers also post comments on Twitter, Instagram, and Facebook. It to desirable to analyze these comments as well.

4.2 Analyzing the Assessment of the Students' Initial Knowledge Level

Once a person is willing to learn the corresponding techniques and ideas, we can start the actual learning. For this learning to be successful, we need to get a good understanding of where the person stands now, what is his/her level of knowledge in the corresponding areas.

To find the current level of a student's knowledge, a sequence of problems of increasing complexity if normally used; if a student can solve a problem, the system generates a more complex one; if a student cannot solve a problem, the system generates an easier one. To find a proper testing scheme, we must take into account that every time a student cannot solve a problem, he/she gets discouraged. To take this into account, in this section, we define an overall effect on a student by combining "positive" and "negative" problems with different weights, and we design a testing scheme which minimizes this effect.

Results from this section first appeared in [25, 26].

Need for a placement test. Computers enable us to provide individualized learning, at a pace tailored to each student. In order to start the learning process, it is important to find out the current level of the student's knowledge, i.e., to place the student at an appropriate level.

Usually, such placement tests use a sequence of N problems of increasing complexity; if a student is able to solve a problem, the system generates a more complex one; if a student cannot solve a problem, the system generates an easier one—until we find the exact level of this student. After this, the actual learning starts.

A seemingly natural idea. A natural tendency is to speed up this preliminary stage and to get to actual leaning as soon as possible, i.e., to minimize the number of problems given to a student.

Resulting solution: bisection. The solution to the corresponding optimization problem is a well-known bisection procedure; see, e.g., [27]. To describe this procedure, let us add, to the problems of levels 1 though N, two fictitious "problems":

- a trivial problem that everyone can solve—which will be called level 0; and
- a very complex problem that no one can solve—which will be called level $N + 1$.

In the beginning, we know that a student can solve a problem at level 0 (since everyone can solve a problem at this level) and cannot solve a problem of level $N + 1$ (since no one can solve problems at this level).

After the tests, we may know that a student can or cannot solve some problems. Let i be the highest level of problems that a student has solved, and let j be the lowest level of problems that a student cannot solve. If $j = i + 1$, then we know exactly where the student stands: he or she can solve problems of level i but cannot solve problems of the next complexity level $i + 1$.

If $j > i = 1$, we need further testing to find out the exact level of knowledge of this student. In the bisection method, we give the student a problem on level

$m \stackrel{\text{def}}{=} (i+j)/2$. Depending on whether a student succeeded in solving this problem or not, we either increase i to m or decrease j to m.

In both cases, we decrease the interval by half. We started with an interval

$$[0, N+1].$$

After s steps, we get an interval of width $2^{-s} \cdot (N+1)$. Thus, when $2^{-s} \cdot (N+1) \le 1$, we get an interval of width 1, i.e., we have determined the student's level of knowledge. This requires $s = \lceil \log_2(N+1) \rceil$ steps.

The problem with bisection. The problem with bisection is that every time a student is unable to solve a problem, he/she gets discouraged; in other words, such problems have a larger effect on the student than problems which the student can solve. For example, if a student is unable to solve any problem already on level 1, this students will get a negative feedback on all $\approx \log_2(N+1)$ problems—and will be thus severely discouraged.

How to solve this problem: an idea. To take the possible discouragement into account, let us define an overall effect on a student by combining "positive" and "negative" problems with different weights.

In other words, we will count an effect of a positive answer as one, and the effect of a negative answer as $w > 1$. For positive answers, the student simply gets tired, while for negative answers, the student also gets stressed and frustrated. The value w can be determined for each individual student.

Resulting optimization problem. For each testing scheme, the resulting effect on each student can be computed as the number of problems that this student solved plus w multiplied by the number of problems that this student did not solve. This effect depends on a student: for some students it may be smaller, for other students it may be larger. As a measure of quality of a testing scheme, let us consider the worst-case effect, i.e., the largest effect over all possible students.

Our objective is to find a testing scheme which places all the students while leading to the smallest effect on a student, i.e., for which the worst-case effect is the smallest possible.

Testing scheme: a general description. A general testing scheme works as follows. First, we ask a student to select a problem of some level n. Depending on whether a student succeeds or not, we ask the student to solve a problem of some other level $n' > n$ or $n' < n$, etc.

As a result, we get the knowledge level of a student, i.e., we get the level i for which the student can solve the problems on this level but cannot solve problems on the next level $i+1$. This level i can take any of the $N+1$ values from 0 to N.

Deriving the main formula. Let $e(x)$ denote the smallest possible effect needed to find out the knowledge level of a student in a situation with $x = N+1$ possible student levels.

In the beginning, we know that a student's level is somewhere between 0 and N. In the optimal testing scheme, we first ask a student to solve a problem of some level

n. Let us consider both possible cases: when the student succeeds in solving this problem and when the student doesn't.

If the student successfully solved the level n problem, this means that after providing a 1 unit of effect on the student, we know that this student's level is somewhere between n and N. In this case, we must select among $N - n + 1 = x - n$ possible student levels. By definition of the function $e(x)$, the remaining effect is equal to $e(x - n)$. Thus, in this case, the total effect on a student is equal to $1 + e(x - n)$.

If the student did not solve the problem of level n, this means that after producing w units of effect on the student, we learn that the student's level is somewhere between 0 and $n - 1$. The remaining effect to determine the student's level is equal to $e(n)$. Thus, the total effect on the student is equal to $w + e(n)$.

The worst-case effect $e(x)$ is, by definition, the largest of the two effects $1+e(x-n)$ and $w + e(n)$: $e(x) = \max(1 + e(x - n), w + e(n))$. In the optimal method, we select n (from 1 to $N = x - 1$) for which this value is the smallest possible. Thus, we conclude that

$$e(x) = \min_{1 \le n < x} \max(1 + e(x - n), w + e(n)). \tag{4.2.1}$$

The value $n(x)$ corresponding to x can be determined as the value for which the right-hand side of the expression (4.2.1) attains its minimum.

Comment. It is worth mentioning that similar formulas appear in other situations; see, e.g., [28, 29]. Because of this similarity, in this section, we have used—after a proper modification—some of the mathematics from [28, 29].

Towards the optimal testing scheme. For $x = 1$, i.e., for $N = 0$, we have $e(1) = 0$. We can use the formula (4.2.1) to sequentially compute the values $e(2)$, $e(3)$, ..., $e(N + 1)$ by using formula (4.2.1); while computing these values, we also compute the corresponding minimizing values $n(2)$, $n(3)$, ..., $n(N + 1)$.

In the beginning, we know that a student's level ℓ is between 0 and N, i.e., that $0 \le \ell < N + 1$. At each stage of the testing scheme, we know that the student's level ℓ is between some numbers i and j: $i \le \ell < j$, where i is the largest of the levels for which the student succeeded in solving the problem, and j is the smallest level for which the student was unable to solve the corresponding problem. In this case, we have $j - i$ possible levels $i, i + 1, \ldots, j - 1$. In accordance with the above algorithm, we should thus ask a question corresponding to the $n(j - i)$-th of these levels. If we count from 0, this means the level $i + n(j - i)$. Thus, we arrive at the following algorithm.

Resulting optimal testing scheme. First, we take $e(1) = 0$, and sequentially compute the values $e(2)$, $e(3)$, ..., $e(N + 1)$ by using the main formula (1), while simultaneously recording the corresponding minimizing values $n(2)$, ..., $n(N + 1)$.

At each stage of testing, we keep track of the bounds i and j for the student's level. In the beginning, $i = 0$ and $j = N + 1$. At each stage, we ask the student to solve a problem at level $m = i + n(j - i)$.

- If the student succeeds in solving this problem, we replace the original lower bound i with the new bound m.
- If the student did not succeed in solving the problem on level m, we replace the original upper bound j with the new bound m.

We stop when $j = i + 1$; this means that the student's level is i.

Example 1. Let us consider an example when $N = 3$ and $w = 3$. In this example, we need to compute the values $e(2)$, $e(3)$, and $e(4)$.

- We take $e(1) = 0$.
- When $x = 2$, the only possible value for n is $n = 1$, so

$$e(2) = \min_{1 \le n < 2} \{\max\{1 + e(2 - n), 3 + e(n)\}\} =$$

$$\max\{1 + e(1), 3 + e(1)\} = \max\{1, 3\} = 3.$$

  Here, $e(2) = 3$, and $n(2) = 1$.
- To find $e(3)$, we must compare two different values $n = 1$ and $n = 2$:

$$e(3) = \min_{1 \le n < 3} \{\max\{1 + e(3 - n)), 3 + e(n)\}\} =$$

$$\min\{\max\{1 + e(2), 3 + e(1)\}, \max\{1 + e(1), 3 + e(2)\}\} =$$

$$\min\{\max\{4, 3\}, \max\{1, 6\}\} = \min\{4, 6\} = 4.$$

  Here, the minimum is attained when $n = 1$, so $n(3) = 1$.
- To find $e(4)$, we must consider three possible values $n = 1$, $n = 2$, and $n = 3$, so

$$e(4) = \min_{1 \le n < 4} \{\max\{1 + e(4 - n), 3 + e(n))\}\} =$$

$$\min\{\max\{1 + e(3), 3 + e(1)\}, \max\{1 + e(2), 3 + e(2)\},$$

$$\max\{1 + e(1), 3 + e(3)\}\} =$$

$$\min\{\max\{5, 3\}, \max\{4, 6\}, \max\{1, 7\}\} =$$

$$\min\{5, 6, 7\} = 5.$$

  Here, the minimum is attained when $n = 1$, so $n(4) = 1$.

So here, the optimal testing procedure is as follows. First, we have $i = 0$ and $j = N+1 = 4$, so we ask a student to solve a problem of level $m = i+n(j-i) = 1$.

If a student did not succeed in solving this level 1 problem, we replace the original upper bound j with the new value $j = 1$. Now, $j = i + 1$, so we conclude that the student is at level 0.

If the student succeeds in solving the level 1 problem, we take $i = 1$ (and keep $j = 4$ the same). In this case, the next problem is of level $m = i + n(j - i) = 2$.

If the student fails to solve the level 2 problem, then we replace the original upper bound j with the new value $j = m = 2$. Here, $j = i + 1$, so we conclude that the student is at level 1.

If the student succeeds is solving the problem at level 2, then we replace the previous lower bound i with the new bound $i = m = 2$. Now, we give the student the next problem of level $i + n(j - i) = 2 + n(4 - 2) = 2 + 1 = 3$.

If the student fails to solve this problem, then we replace the original upper bound j with the new value $j = m = 3$. Here, $j = i + 1$, so we conclude that the student is at level 2.

If the student succeeds in solving the problem at level 3, then we replace the previous lower bound i with the new bound $i = m = 3$. Here, $j = i + 1$, so we conclude that the student is at level 2.

Comment. In this case, the optimal testing scheme is the most cautious one, when we increase the level by one every time. This way, we are guaranteed that a tested student has no more than one negative experience.

Example 2. Let us now consider an example when $N = 3$ and $w = 1.5$.

- We take $e(1) = 0$.
- When $x = 2$, then

$$e(2) = \min_{1 \le n < 2} \{\max\{1 + e(2 - n), 3 + e(n)\}\} =$$

$$\max\{1 + e(1), 1.5 + e(1)\} = \max\{1, 1.5\} = 1.5.$$

  Here, $e(2) = 1.5$, and $n(2) = 1$.
- To find $e(3)$, we must compare two different values $n = 1$ and $n = 2$:

$$e(3) = \min_{1 \le n < 3} \{\max\{1 + e(3 - n)), 1.5 + e(n)\}\} =$$

$$\min\{\max\{1 + e(2), 1.5 + e(1)\}, \max\{1 + e(1), 1.5 + e(2)\}\} =$$

$$\min\{\max\{2.5, 1.5\}, \max\{1, 3\}\} = \min\{2.5, 3\} = 2.5.$$

  Here, the minimum is attained when $n = 1$, so $n(3) = 1$.
- To find $e(4)$, we must consider three possible values $n = 1$, $n = 2$, and $n = 3$, so

$$e(4) = \min_{1 \le n < 4} \{\max\{1 + e(4 - n), 1.5 + e(n))\}\} =$$

$$\min\{\max\{1 + e(3), 1.5 + e(1)\}, \max\{1 + e(2), 1.5 + e(2)\},$$

$$\max\{1 + e(1), 1.5 + e(3)\}\} =$$

$$\min\{\max\{3.5, 1.5\}, \max\{2.5, 3\}, \max\{1, 4\}\} =$$

$$\min\{3.5, 3, 4\} = 3.$$

Here, the minimum is attained when $n = 2$, so $n(4) = 2$.

So here, the optimal testing procedure is as follows. First, we have $i = 0$ and $j = N+1 = 4$, so we ask a student to solve a problem of level $m = i+n(j-i) = 2$.

If a student did not succeed in solving the level 2 problem, we replacing the original upper bound j with the new value $j = 2$. Now, we ask the student to solve a problem on level $m = i + n(j - i) = 1$. If a student succeeds, his/her level is 1; if the student does not succeed, his/her level is 0.

If the student succeeds in solving the level 2 problem, we take $i = 2$ (and keep $j = 4$ the same). In this case, the next problem is of level $m = i + n(j - i) = 3$. If a student succeeds, his/her level is 3; if the student does not succeed, his/her level is 2.

Comment. In this case, the optimal testing scheme is the bisection.

Computational complexity. For each n from 1 to N, we need to compare n different values. So, the total number of computational steps is proportional to $1+2+\ldots+N = O(N^2)$.

Additional problem. When N is large, N^2 may be too large. In some applications, the computation of the optimal testing scheme may takes too long. For this case, we have developed a faster algorithm for producing a testing scheme which is only asymptotically optimal.

A faster algorithm for generating an asymptotically optimal testing scheme: description. First, we find the real number $\alpha \in [0, 1]$ for which $\alpha + \alpha^w = 1$. This value α can be obtained, e.g., by applying bisection [27] to the equation $\alpha + \alpha^w = 1$.

Then, at each step, once we have the lower bound i and the upper bound j for the (unknown) student level ℓ, we ask the student to solve a problem at the level

$$m = \lfloor \alpha \cdot i + (1 - \alpha) \cdot j \rfloor.$$

Comments. This algorithm is similar to bisection, expect that bisection corresponds to $\alpha = 0.5$. This makes sense, since for $w = 1$, the equation for α takes the form $2\alpha = 1$, hence $\alpha = 0.5$. For $w = 2$, the solution to the equation $\alpha + \alpha^2 = 1$ is the well-known golden ratio $\alpha = \dfrac{\sqrt{5} - 1}{2} \approx 0.618$.

Computational complexity. At each step, we end up with either an interval $[i, m]$ whose width is $1 - \alpha$ from the original size, or with the interval $[m, j]$ whose width is α from the original size. Since $\alpha \geq 1 - \alpha$, the worst-case decrease is decrease by a factor of α. In k steps, we decrease the width N to $\leq N \cdot \alpha^k$. Thus, we stop for sure when $N \cdot \alpha^k \leq 1$, i.e., after $k = O(\log(N))$ problems.

At each level, we need a constant number of computation steps to compute the next level, so the overall computation time is $O(\log(N))$.

In what sense the resulting testing scheme is asymptotically optimal. We will prove that for this scheme, there is a constant C such that for every N, the worst-case effect from this scheme differs from the worst-case effect of the optimal testing scheme by no more than C.

Proof that the resulting testing scheme is indeed asymptotically optimal. Let us denote the optimal effect by $e(N)$ and the worst-case effect corresponding to our procedure by $e_0(N)$. Let us also denote $K = 2^{-\alpha}$. To prove our result, we will prove that there exist constants $C > 0$ and $C_1 > 0$ such that for every N, we have

$$K \cdot \log_2(N) \le e(N)$$

and

$$e_0(N) \le K \cdot \log_2(N) + C - \frac{C_1}{N}.$$

By definition, $e(N)$ is the smallest worst-case effect of all possible testing schemes, thus, $e(N) \le e_0(N)$. So, if we prove the above two inequalities, we will indeed prove that our algorithm is asymptotically optimal.

Proof of the first inequality. Let us first prove the first inequality by induction over N. The value $N = 1$ represents the induction base. For this value, $K \cdot \log_2(1) = 0 = e(1)$, so the inequality holds.

Let us now describe the induction step. Suppose that we have already proved the inequality $K \cdot \log_2(n) \le e(n)$ for all $n < N$. Let us prove that $K \cdot \log_2(N) \le e(N)$.

Due to our main formula, $e(N)$ is the smallest of the values

$$\max\{1 + e(x - n), w + e(n)\}$$

over $n = 1, 2, \ldots, N - 1$. So, to prove that $K \cdot \log_2(N)$ is indeed the lower bound for $e(N)$, we must prove that $K \cdot \log_2(N)$ cannot exceed each of these values, i.e., that

$$K \cdot \log_2(N) \le \max\{1 + e(N - n), w + e(n)\}$$

for every $n = 1, 2, \ldots, N - 1$. For these n, we have $n < N$ and $N - n < N$, so for all these values, we already know that $K \cdot \log_2(n) \le e(n)$ and

$$K \cdot \log_2(N - n) \le e(N - n).$$

Therefore,

$$1 + K \cdot \log_2(N - n) \le 1 + e(N - n),$$

$$w + K \cdot \log_2(n) \le w + e(n),$$

and

$$\max\{1 + K \cdot \log_2(N - n), w + K \cdot \log_2(n)\} \le$$

$$\max\{1 + e(N - n), w + e(n)\}.$$

So, to prove the desired inequality, it is sufficient to prove that

$$K \cdot \log_2(N) \le$$

$$\max\{1 + K \cdot \log_2(N - n), w + K \cdot \log_2(n)\}.$$

We will prove this inequality by considering two possible cases: $n \le (1 - \alpha) \cdot N$ and $n \ge (1 - \alpha) \cdot N$.

- When $n \le (1 - \alpha) \cdot N$, we have $N - n \ge \alpha \cdot N$ and therefore,

  $$1 + K \cdot \log_2(N - n) \ge z,$$

  where

  $$z \stackrel{\text{def}}{=} 1 + K \cdot \log_2(\alpha \cdot N) = 1 + K \cdot \log_2(N) + K \cdot \log_2(\alpha).$$

  Here, by definition of $K = 2^{-\alpha}$, we have $\log_2(\alpha) = -1/K$, hence

  $$1 + K \cdot \log_2(\alpha) = 0,$$

  and so $z = K \cdot \log_2(N)$. In this case,

  $$K \cdot \log_2(N) \le z = 1 + K \cdot \log_2(N - n) \le$$

  $$\max\{1 + K \cdot \log_2(N - n), w + K \cdot \log_2(n)\}.$$

- When $n \ge (1 - \alpha) \cdot N$, we have $w + K \cdot \log_2(n) \ge t$, where

  $$t \stackrel{\text{def}}{=} w + K \cdot \log_2((1 - \alpha) \cdot N) = w + K \cdot \log_2(N) + K \cdot \log_2(1 - \alpha).$$

  By definition of α, we have $1 - \alpha = \alpha^w$, so $\log_2(1 - \alpha) = w \cdot \log_2(\alpha)$ and thus, $w + K \cdot \log_2(1 - \alpha) = w \cdot (1 + K \cdot \log_2(\alpha)) = 0$. Hence, $t = K \cdot \log_2(N)$. So, in this case,

  $$K \cdot \log_2(N) \le t = w + K \cdot \log_2(n) \le$$

  $$\max\{1 + \log_2(N - n), w + K \cdot \log_2(n)\}.$$

In both cases, we have the desired inequality. The induction step is proven, and so, indeed, for every N, we have

$$K \cdot \log_2(N) \le e(N).$$

Proof of the second inequality. Let us now prove that there exist real numbers $C > 0$ and $C_1 > 0$ for which, for all N,

$$e_0(N) \le K \cdot \log_2(N) + C - \frac{C_1}{N}.$$

To prove this inequality, we will pick a value N_0, prove that this inequality holds for all $N \le N_0$, and then use mathematical induction to show that it holds for all $N > N_0$ as well.

Induction basis. Let us first find the conditions on C, C_1, and N_0 under which for all $N \le N_0$,

$$e_0(N) \le K \cdot \log_2(N) + C - \frac{C_1}{N}.$$

Subtracting $K \cdot \log_2(N)$ and adding $\frac{C_1}{N}$ to both sides of the this inequality, we get

$$C \ge \frac{C_1}{N} + e_0(N) - K \cdot \log_2(N)$$

for all N from 1 to N_0. So, to guarantee that this inequality holds, if we have already chosen N_0 and C_1, we can choose

$$C = \max_{1 \le N \le N_0} \left(\frac{C_1}{N} + e_0(N) - K \cdot \log_2(N) \right).$$

Induction step. Let us assume that for all $n < N$ (where $N > N_0$), we have proven that

$$e_0(n) \le K \cdot \log_2(n) + C - \frac{C_1}{n}.$$

We would like to conclude that

$$e_0(N) \le K \cdot \log_2(N) + C - \frac{C_1}{N}.$$

According to the definition of $e_0(N)$, we have

$$e_0(N) = \max\{1 + e_0(N - n), w + e_0(n))\},$$

where $n = \lfloor (1 - \alpha) \cdot N \rfloor$. Due to induction hypothesis, we have

$$e_0(n) \le K \cdot \log_2(n) + C - \frac{C_1}{n}$$

and

$$e_0(N-n) \le K \cdot \log_2(N-n) + C - \frac{C_1}{N-n}.$$

Therefore,

$$e_0(N) \le \max \left\{ 1 + K \cdot \log_2(N-n) + C - \frac{C_1}{N-n}, \right.$$

$$\left. w + K \cdot \log_2(n) + C - \frac{C_1}{n} \right\}.$$

Thus, to complete the proof, it is sufficient to conclude that this maximum does not exceed

$$K \cdot \log_2(N) + C - \frac{C_1}{N}.$$

In other words, we must prove that

$$1 + K \cdot \log_2(N-n) + C - \frac{C_1}{N-n} \le K \cdot \log_2(N) + C - \frac{C_1}{N}$$

and that

$$w + K \cdot \log_2(n) + C - \frac{C_1}{N-n} \le K \cdot \log_2(N) + C - \frac{C_1}{N}. \quad (4.2.2)$$

Without losing generality, let us show how we can prove the second of these two inequalities. Since $n = \lfloor (1-\alpha) \cdot N \rfloor$, the left-hand side of the inequality (2) can be rewritten as

$$W_1 + K \cdot \log_2((1-\alpha) \cdot N) + K \cdot \big(\log_2(n) - \log_2((1-\alpha) \cdot N)\big) + C - \frac{C_1}{n}.$$

We already know that $w + K \cdot \log_2((1-\alpha) \cdot N) = K \cdot \log_2(N)$. Thus, the left-hand side of (2) takes the simpler form

$$K \cdot \log_2(N) + K \cdot \big(\log_2(n) - \log_2((1-\alpha) \cdot N)\big) + C - \frac{C_1}{n}.$$

Substituting this expression into (2) and canceling the terms $K \cdot \log_2(N)$ and C in both sides, we get an equivalent inequality

$$K \cdot \big(\log_2(n) - \log_2((1-\alpha) \cdot N)\big) - \frac{C_1}{n} \le -\frac{C_1}{N}. \quad (4.2.3)$$

Let us further simplify this inequality. We will start by estimating the difference $\log_2(n) - \log_2((1-\alpha) \cdot N)$. To estimate this difference, we will use the intermediate value theorem, according to which, for every smooth function $f(x)$, and for arbitrary two values a and b, we have $f(a) - f(b) = (a - b) \cdot f'(\xi)$ for some $\xi \in [a, b]$. In our case,

$$f(x) = \log_2(x) = \frac{\ln(x)}{\ln(2)},$$

$a = n$, and $b = (1 - \alpha) \cdot N$. Here,

$$f'(\xi) = \frac{1}{\xi \cdot \ln(2)},$$

so

$$f'(\xi) \leq \frac{1}{n \cdot \ln(2)};$$

also, $|a - b| \leq 1$, so, the difference $\log_2(n) - \log_2((1 - \alpha) \cdot N)$ can be estimated from above by:

$$\log_2(n) - \log_2((1 - \alpha) \cdot N) \leq \frac{1}{n \cdot \ln(2)}.$$

Hence, the above inequality holds if the following stronger inequality holds:

$$\frac{K}{n \cdot \ln(2)} - \frac{C_1}{n} \leq -\frac{C_1}{N},$$

or, equivalently,

$$\frac{C_1}{N} \leq \frac{C_1 - K/\ln(2)}{n}. \tag{4.2.4}$$

Here, $n \geq (1 - \alpha) \cdot N - 1$, i.e.,

$$\frac{n}{N} \geq (1 - \alpha) - \frac{1}{n}.$$

When $N \to \infty$, we have $n \to \infty$ and $\frac{1}{n} \to 0$. Thus, for every $\varepsilon > 0$, there exists an N_0 starting from which $\frac{1}{n} \leq \varepsilon$ and hence, $n \geq (1 - \alpha - \varepsilon) \cdot N$. For such sufficiently large N, the inequality (4.2.4) can be proven if we have

$$\frac{C_1}{N} \leq \frac{C_1 - K/\ln(2)}{(1 - \alpha - \varepsilon) \cdot N},$$

i.e., if we have

$$C_1 \leq \frac{C_1 - K/\ln(2)}{1 - \alpha - \varepsilon}. \tag{4.2.5}$$

Since $0 \leq \alpha \leq 1$, for sufficiently large C_1, this inequality is true. For such C_1, therefore, the induction can be proven and thus, the second inequality is true.

The statement is proven.

What if we also know probabilities: formulation of the problem. In some cases, we also know the frequencies $p_0, p_1, \ldots, p_N$ with which students are at the corresponding levels. These frequencies can be alternatively described by the corresponding cumulative distribution function $F(i) \stackrel{\text{def}}{=} \text{Prob}(\ell < i) = p_0 + p_1 + \ldots + p_{i-1}$. In this situation, instead of the *worst-case* effect, we can alternatively consider the *average* effect—and look for a testing scheme which minimizes the average effect.

Towards a scheme which minimizes the average effect. Let $e(i, j)$ be the smallest possible conditional average effect under the condition that a student's actual level is between i and j, i.e., that the student has successfully solves a problem at level i and was unable to solve the problem at level j. (The original situation corresponds to $i = 0$ and $j = N + 1$.)

In this situation, we ask the student to solve a problem at some level $n \in (i, j)$. Let us consider both possible cases: when the student was able to solve this problem, and when a student was unable to solve this problem.

If a student was able to solve the problem at level n, this means that the student's level is in between n and j. By definition of a function $e(\cdot, \cdot)$, the expected remaining effect is equal to $e(n, j)$. Thus, in this case, the overall expected effect on the student is equal to $1 + e(n, j)$. The conditional probability of this case can be obtained by dividing the probability $F(j) - F(n)$ that the student's level is between n and j by the original probability $F(j) - f(i)$ that the student's level is between i and j. Thus, this probability is equal to $\frac{F(j) - F(n)}{F(j) - F(i)}$.

If a student was unable to solve the problem at level n, this means that the student's level is in between i and n. By definition of a function $e(\cdot, \cdot)$, the expected remaining effect is equal to $e(i, n)$. Thus, in this case, the overall expected effect on the student is equal to $w + e(i, n)$. The conditional probability of this case can be obtained by dividing the probability $F(n) - F(i)$ that the student's level is between i and n by the original probability $F(j) - f(i)$ that the student's level is between i and j. Thus, this probability is equal to $\frac{F(n) - F(i)}{F(j) - F(i)}$.

Thus, we have the expected effect $1 + e(n, j)$ with probability $\frac{F(j) - F(n)}{F(j) - F(i)}$, and the expected effect $w + e(i, n)$ with probability $\frac{F(n) - F(i)}{F(j) - F(i)}$. So, the overall expected effect is equal to

$$\frac{F(j) - F(n)}{F(j) - F(i)} \cdot (1 + e(n, j)) + \frac{F(n) - F(i)}{F(j) - F(i)} \cdot (w + e(i, n)).$$

Since we want to minimize the average effect, we select n for which this value is the smallest possible. Thus, we arrive at the following formula:

Main formula: average case.

$$e(i, j) =$$

$$\min_{i \le n < j} \left(\frac{F(j) - F(n)}{F(j) - F(i)} \cdot (1 + e(n, j)) + \frac{F(n) - F(i)}{F(j) - F(i)} \cdot (w + e(i, n)) \right). \quad (4.2.6)$$

Towards the optimal testing scheme. When $j = i + 1$, we know that the student's level is i, so no additional testing is needed and the effect is 0: $e(i, i + 1) = 0$. We can start with these values and sequentially use the formula (4.2.6) to compute the values $e(i, i+2)$, $e(i, i+3)$, etc. In each case, we find $n(i, j)$ for which the minimum is attained.

Resulting optimal testing scheme. First, we take $e(i, i + 1) = 0$ for all i, and use the formula (4.2.6) to sequentially compute the values $e(i, i+2)$, $e(i, i+3)$, ..., until we cover all possible values $e(i, j)$. For each i and j, we record the value $n(i, j)$ for which the right-hand side of the formula (4.2.6) is the smallest.

At each stage of the testing, we keep track of the bounds i and j for the student's level. In the beginning, $i = 0$ and $j = N + 1$. At each stage, we ask the student to solve a problem at level $m = n(i, j)$.

- If the student succeeds in solving this problem, we replace the original lower bound i with the new bound m.
- If the student did not succeed in solving the problem on level m, we replace the original upper bound j with the new bound j.

We stop when $j = i + 1$; this means that the student's level is i.

Computational complexity. For each of $O(N^2)$ pairs $i < j$ of numbers from 0 to N, we need to compare $j - i = O(N)$ different values. So, the total number of computational steps is proportional to $O(N^2) \cdot O(N) = O(N^3)$.
Comment. For large N, this computation time may be too large. It would be nice—similarly to the worst-case optimization—to come up with a faster algorithm even if it generates only an asymptotically optimal testing scheme.

4.3 Analyzing the Way the Material Is Presented: Global Aspects

Once this information is known, we need to actually present this information to the interested folks—and use appropriate feedback to modify (if needed) the speed with which this knowledge is presented. Issues related to the material's presentation are analyzed in this section and in the following section. In the two subsections that form this section, we consider the problem from the global viewpoint: in what order we should present different parts of the material, and how individualized the presentation should be—should we present the exact same material to all the students, or should we allow students to select among several possible versions of the presentations.

4.3.1 In What Order Should We Present the Material

In the traditional approach to learning, if we want students to learn how to solve different types of problems, we first teach them how to solve problems of the first type, then how to solve problems of the second type, etc. It turns out that we can speed up learning if we interleave problems of different types. In particular, it has bene empirically shown that interleaving problems of four different types leads to a double speed-up. In this section, we provide a possible explanation for this empirical fact.

The results from this section were first published in [30].

Traditional approach to learning several skills. Traditionally, when students need to learn several skills, they learn them one by one:

- first, they learn the skill a;
- once they have mastered skill a, they start learning skill b,
- etc.

For example, in a geometry class, students need to learn how to solve several different types of problems. For that purpose:

- they first spend several class periods learning how to solve problems of type a,
- then they spend several class periods learning how to solve problems of type b,
- etc.

Interleaving: an alternative approach. An alternative approach is *interleaving*, when students learn several skills at the same time. For example, instead of first solving several problems of type a, then several problem of type b, etc., they solve a problem of type a, then a problem of type b, etc., then again a problem of type a, then again b, etc.

In other words, instead of a usual sequence of problem types

$$aaa \ldots bbb \ldots ccc \ldots$$

we use an interleaving sequence

$$abc \ldots abc \ldots abc \ldots$$

Interleaving enhances learning. Several studies show that interleaving enhances different types of learning, from learning to play basketball [31, 32] to learning art [33] to learning mathematics [34–36]; see also [37].

Quantitative fact. In particular, in [36], it is shown that interleaving of four different types of geometric problems increases the average number of correct answers on the test twice, from 38 to 77%.

In other words, interleaving of four different types of problems doubles the learning speed.

What we do in this section. In this section, we provide a possible explanation to the above enhancement, an explanation based on a simple geometric model.

A simple geometric model. Let us describe traditional and interleaved approaches in geometric terms. We want students to learn to solve four different types of problems. In the beginning, the students do not know how to solve any of these problems. The objective is for them to be able to solve all four types of problems.

We can represent the state of the students at each moment of time by the percentage (x_1, x_2, x_3, x_4) of problems of each type that a student can solve.

- In the beginning, the students are in the state $(0, 0, 0, 0)$.
- Our objective is to reach the state $(1, 1, 1, 1)$.

How traditional approach is represented in this geometric model. In the traditional approach, the students first learn to solve problems of the first type, then they learn how to solve problems of the second type, etc. In other words:

- the students first move from the state $(0, 0, 0, 0)$ to the state $(1, 0, 0, 0)$,
- then they move to the state $(1, 1, 0, 0)$,
- after that, they move to the state $(1, 1, 1, 0)$,
- and, finally, they move to the desired state $(1, 1, 1, 1)$.

At each stage of this process, we can assume that the students follow the shortest path—a straight line—to get to the corresponding state.

Each stage has length 1, so the total length of all four stages is equal to 4.

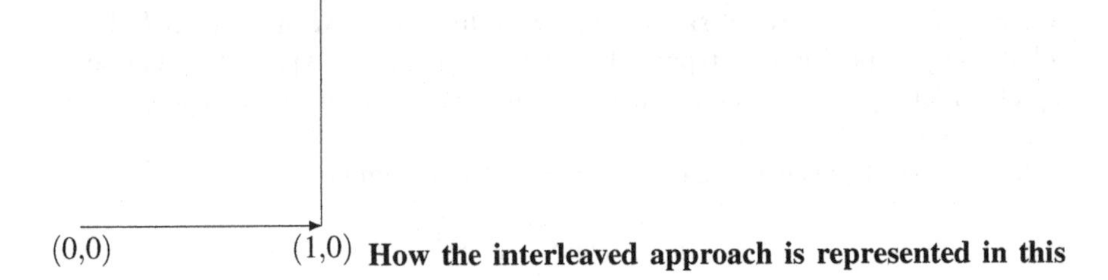

How the interleaved approach is represented in this geometric model. In the interleaved approach, at each moment of time, the students have spent equal time on problems of all fours types and thus, their skills in solving problems of all four types are equal.

In geometric terms, this means that their state is described by a tuple (x, x, x, x). Thus, for this approach, learning follows the diagonal path

$$\{(x, x, x, x) : x \in [0, 1]\}.$$

This diagonal is the straight line segment connecting the original state $(0, 0, 0, 0)$ with the desired state $(1, 1, 1, 1)$.

The length of this path is equal to the distance between these two states $(0, 0, 0, 0)$ and $(1, 1, 1, 1)$, i.e., to the value

$$\sqrt{(1-0)^2+(1-0)^2+(1-0)^2+(1-0)^2} = \sqrt{4} = 2.$$

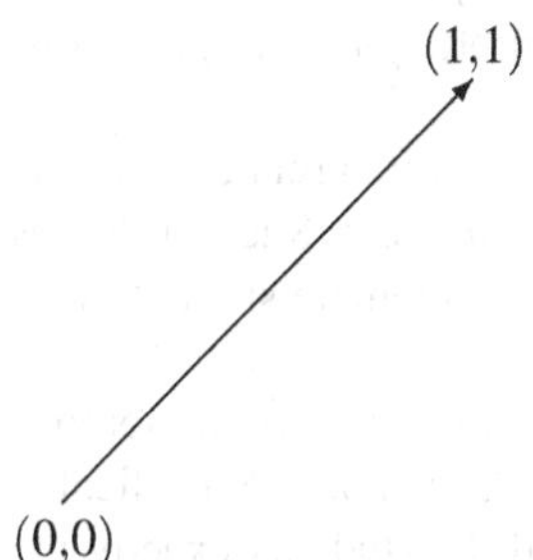

Resulting explanation of the empirical fact. We see that in the interleaved approach, the path to the desired state is *twice shorter* than in the traditional approach. This may explain why, when we interleave four different types of problems, learning becomes *twice faster*.

4.3.2 How Much Flexibility Should We Allow Students

In many pedagogical situations, it is advantageous to give students some autonomy: for example, instead of assigning the same homework problem to all the students, to give students a choice between several similar problems, so that each student can choose a problem whose context best fits his or her experiences. A recent experimental study shows that there is a 45% correlation between degree of autonomy and student success. In this section, we provide a theoretical explanation for this correlation value.

Results from this section first appeared in [38].

Traditional non-autonomous approach to learning. Traditionally, all the students in the class are assigned the same homework problems, the same practice problems, and are given the same problems on the tests.

Need for autonomy. The main advantage of the traditional approach seems to be that all the students are treated equally: they are given exactly the same problems, so they have the same chance to succeed.

In reality, however, the traditional approach has its limitations. For example, in a university setting, every engineering student has to take calculus. Engineering students usually form a majority in calculus classes. However, in addition to engineering students, also students from other disciplines, e.g., from bioinformatics, are required to take calculus. Since the majority of students in a calculus class are engineering students, most textbook application-related problems come from engineering or from related fields. This fact gives an unfair disadvantage to bioinformatics students many of whom are not very familiar with the main concepts from engineering.

As a result, it is much harder for a bioinformatics students to study in this class, and for those bioinformatics students who worked hard and mastered all the concepts, it is more difficult to show their knowledge on the tests—since the problems given on these tests are also skewed towards engineering-related topics.

This problem will not disappear if we simply switch to biology-related problems: then, it will be unfair to engineering students.

Similarly, in middle and high school, physics problems are often related to activities familiar to kids, such as soccer, basketball, skateboarding, etc. Such problems enliven the class but they give an unfair disadvantage, e.g., to female students who are usually less involved in soccer than boys.

In all these cases, to eliminate the unfairness of the traditional non-autonomous approach, it is desirable to give students some degree of *autonomy*. Specifically, on each assignment and on each test, instead of giving all the students exactly the same problem corresponding to a certain topic, it is desirable to give students *several* problems to choose from, so that each student will be able to decide which of the problem he or she wants to solve.

There is also an additional psychological advantage of providing students with autonomy: since the students themselves have to make choices, they feel more in control of the learning process, and it is well known that people usually perform better when they are (at least partly) in control than when they simply blindly follow others' instructions.

Autonomy indeed improves learning: an empirical fact. The need for student autonomy is well understood by many teachers, and many teachers have added elements of autonomy to their teaching. There is a large amount of anecdotal evidence showing that autonomy improves learning. Recently, this improvement was confirmed by a rigorous study [39] that showed that there is indeed a high correlation (45%) between the degree of autonomy and the students' success.

The empirical data needs a theoretical explanation: what we do in this section. While the empirical data is very convincing, it is always desirable to come up with a theoretical explanation for this data. To be more precise, the paper [39] provides a deep *qualitative* theoretical explanation of its results. It is desirable to transform this qualitative explanation into a *quantitative* one, i.e., into an explanation that would not only explain the positiveness of the correlation, but that would also help us predict the numerical value of this correlation. Such an explanation is provided in this section.

Towards a mathematical model. To provide the desired quantitative explanation, let us formulate a simple mathematical model for autonomy.

As we have mentioned earlier, the main reason why we need autonomy is that for the same class of problems, at the same level of student understanding, the students will show different degree of success depending on how familiar they are with the overall context of this problem.

In other words, in different contexts, the students will exhibit different degrees of success x (e.g., grades). Let us fix a specific situation, i.e., a specific topic and a specific level of understanding of this topic. Let us denote, by $\underline{x}$, the smallest of these degrees corresponding to this level of understanding, and by $\overline{x}$, the largest of these degrees. Thus, depending on the context in which we present the corresponding problem, the grade of the same student may take any value from $\underline{x}$ to $\overline{x}$.

How to estimate probabilities of different values from $\underline{x}$ to $\overline{x}$? To properly gauge the effect of autonomy on student's learning, we should know the probability of different values x from the interval $[\underline{x}, \overline{x}]$.

In general, we have no reason to believe that some values from this interval are more frequent than others. So, it is reasonable to assume that all the values from this interval are equally probable, i.e., that we have a uniform distribution on this interval; see, e.g., [40].

How to describe case of autonomy. Let us now describe the probability distribution corresponding to the autonomy case.

Let us assume that, instead of single problem, we are given the student a choice between k different problems. For each of these problems, the success rate x_i ($i = 1, \ldots, k$), the success rate is uniformly distributed in the interval $[\underline{x}, \overline{x}]$. Out of these k problems, the student will choose the one with which context he or she is most familiar, i.e., the one with the largest expected success rate $m_k \stackrel{\text{def}}{=} \max(x_1, \ldots, x_k)$. Thus, in the case of autonomy, the resulting success rate is distributed as the maximum of k independent random variables each of which is uniformly distributed on the given interval.

How many alternative problems should we design? On the one hand, the more choices, the better. On the other hand, good problems are not easy to design, and coming up with many additional problems would be very time-consuming. Let us therefore stop when the further increase in student success is not statistically significant.

Usually, in applications of statistics, a 5% threshold is used to describe statistical significance; see, e.g., [17]. So, we will stop when the difference between the expected grade $E[m_k]$ corresponding to k problems and the expected grade $E[m_{k+1}]$ (which will occurs if we add one more problem) does not exceed 5%.

For each k and for each x, the maximum m_k of k values x_i is smaller than or equal to x if and only if each of these values is $\le x$. Thus, due to independence assumption,

$$F_k(x) \stackrel{\text{def}}{=} \text{Prob}(m_k \le x) = \text{Prob}((x_1 \le x) \,\&\, \ldots \,\&\, (x_k \le x)) =$$

$$\text{Prob}(x_1 \le x) \cdot \ldots \cdot \text{Prob}(x_k \le x).$$

For the uniform distribution, $\text{Prob}(x_i \le x) = \dfrac{x - \underline{x}}{\overline{x} - \underline{x}}$, so $F_k(x) = \left(\dfrac{x - \underline{x}}{\overline{x} - \underline{x}}\right)^k$. Thus, the corresponding probability density function $f_k(x)$ has the form

$$f_k(x) = \frac{dF_k(x)}{dx} = k \cdot \frac{(x - \underline{x})^{k-1}}{(\overline{x} - \underline{x})^k}.$$

Therefore, the mean grade $E[m_k]$ is equal to

$$E[m_k] = \int_{\underline{x}}^{\overline{x}} x \cdot f_k(x)\,dx = \int_{\underline{x}}^{\overline{x}} x \cdot k \cdot \frac{(x-\underline{x})^{k-1}}{(\overline{x}-\underline{x})^k}\,dx.$$

By introducing a new variable $x' \stackrel{\text{def}}{=} x - \underline{x}$, for which $x = \underline{x} + x'$, we can explicitly compute the corresponding integral, and get $E[m_k] = \underline{x} + \frac{k}{k+1} \cdot (\overline{x} - \underline{x})$. So, the $E[m_k]$ is at level $\frac{k}{k+1}$ in the interval $[\underline{x}, \overline{x}]$.

For $k = 1$, we get $\frac{1}{2} = 50\%$ of this interval. For $k = 2$, we get $\frac{2}{3} \approx 67\%$ of this interval—a statistically significantly larger value, since $67 - 50 > 5$. For $k = 3$, we get $\frac{3}{4} = 75\%$, which is also statistically significantly larger value ($75 - 67 > 5$). For $k = 4$, we get $\frac{4}{5} = 80\%$ which is *not* statistically significantly larger value, since $80 - 75 = 5$.

Thus, we select $k = 3$ alternatives for each problem. In this case, the probability distribution for the success rate m_3 can be described by the probability density

$$f_3(x) = k \cdot \frac{(x-\underline{x})^2}{(\overline{x}-\underline{x})^3}.$$

Let us now compute the correlation. We have described the corresponding mathematical model. Let us now use this model to compute the correlation between the student's success and the autonomy level.

What we want to estimate. We want to find the correlation ρ between the success rate X and the autonomy level Y. In general, the correlation has the form

$$\rho = \frac{E[X \cdot Y] - E[X] \cdot E[Y]}{\sigma[X] \cdot \sigma[Y]}, \tag{4.3.1}$$

where $\sigma = \sqrt{V}$ is the standard deviation of the corresponding random variable.

How do we describe Y. Here, we only consider two levels of autonomy: no autonomy and giving a student the maximum choice (of 3 problems). Without losing generality, let us denote the autonomy case by $Y = 1$, and the non-autonomy case by $Y = -1$.

When researchers experimentally compare two techniques, they random assign each technique to some objects (in this case, to classes). To make comparison maximally fair, it is desirable to treat both techniques equally, in particular, to assign the same number of objects to each technique. In this case, we get $Y = 1$ and $Y = -1$ with the same probability 0.5. Thus, $E[Y] = 0$, so $V[Y] = E[(Y - E[Y])^2] = E[Y^2] = 1$ and $\sigma[Y] = \sqrt{V[Y]} = 1$.

Re-scaling X. It is known that the correlation does not change if we linearly re-scale each quantity x, i.e., replace it with $x' = a \cdot X + b$. For example, the correlation between height and weight should be the same whether we use inches and pounds or centimeters and kilograms.

We can use this fact to replace the original value $x \in [\underline{x}, \overline{x}]$ with an easier-to-analyze value $X = \dfrac{x - \underline{x}}{\overline{x} - \underline{x}} \in [0, 1]$. In this case, X is uniformly distributed on the interval $[0, 1]$ when $Y = 1$ and distributed as m_3, with probability density function $f_3(x) = 3x^2$, when $Y = -1$.

Estimating $E[X \cdot Y]$. Since both values $Y = \pm 1$ occur with probability $\dfrac{1}{2}$, we get:

$$E[X \cdot Y] = \frac{1}{2} \cdot E[X \cdot Y \mid Y = 1] + \frac{1}{2} \cdot E[X \cdot Y \mid Y = -1] =$$

$$\frac{1}{2} \cdot E[X \mid Y = 1] - \frac{1}{2} \cdot E[X \mid Y = -1].$$

For the uniform distribution, $E[X \mid Y = -1] = \dfrac{1}{2}$, and for the distribution m_3, as we have mentioned earlier, $E(X \mid Y = 1] = \dfrac{3}{4}$, so

$$E[X \cdot Y] = \frac{1}{2} \cdot \frac{3}{4} - \frac{1}{2} \cdot \frac{1}{2} = \frac{1}{8}. \quad (4.3.2)$$

Estimating $\sigma[X]$. Similarly, for the variance $V[X] = E[X^2] - (E[X])^2$, we have

$$E[X] = \frac{1}{2} \cdot E[X \mid Y = 1] + \frac{1}{2} \cdot E[X \mid Y = -1] = \frac{1}{2} \cdot \frac{3}{4} + \frac{1}{2} \cdot \frac{1}{2} = \frac{5}{8},$$

and $E[X^2] = \dfrac{1}{2} \cdot E[X^2 \mid Y = 1] + \dfrac{1}{2} \cdot E[X^2 \mid Y = -1]$. For the uniform distribution, with uniform probability density $f_1(x) = 1$, we have

$$E[X^2 \mid Y = -1] = \int_0^1 x^2 \cdot f_1(x)\, dx = \frac{1}{3}.$$

Similarly,

$$E[X^2 \mid Y = 1] = \int_0^1 x^2 \cdot f_3(x)\, dx = \int_0^2 x^2 \cdot 3x^2\, dx = \frac{3}{5}.$$

Thus,

$$E[X^2] = \frac{1}{2} \cdot \frac{1}{3} + \frac{1}{2} \cdot \frac{3}{5} = \frac{7}{15}.$$

Hence,

$$V[X] = E[X^2] - (E[X])^2 = \frac{7}{15} - \left(\frac{5}{8}\right)^2 = \frac{7}{15} - \frac{25}{64} = \frac{73}{15 \cdot 64}.$$

Hence,

$$\sigma[X] = \sqrt{V[X]} = \sqrt{\frac{73}{15}} \cdot \frac{1}{8}. \tag{4.3.3}$$

Resulting estimate. Substituting the expressions $E[Y] = 0$, $\sigma[Y] = 1$, (4.3.2), and (4.3.3) into the formula (4.3.1), we get

$$\rho = \frac{\frac{1}{8}}{\sqrt{\frac{73}{15}} \cdot \frac{1}{8}} = \sqrt{\frac{15}{73}} \approx 45\%.$$

Thus, we get a theoretical explanation for the empirical correlation.

4.4 Analyzing the Way the Material Is Presented: Local Aspects

In this section, we continue the analysis of the problems related to the material's presentation. In the previous section, we considered the problem from the global viewpoint: e.g., in what order we should present different parts of the material. In this section, we consider this problem from the local viewpoint: what is the best way to present different items.

As we mentioned earlier, these issues are especially important in computerized education, when the material is presented by an automated system, without a detailed supervision from a human teacher.

To design such systems, practitioners use empirical rules and laws derived from the experience of computerized information access. One of the most well-known of these laws is an empirical Fitts's Law, according to which the average time T of accessing an icon of size w at a distance d from the center of the screen is proportional to the logarithm of the ratio w/d. There exist explanations for this law, but these explanations have gaps.

In this section, we show that these gaps can be explained if we analyze this problem from the geometric viewpoint. Thus, we get a possible theoretical explanation of the Fitts's Law.

What is Fitts's Law. The efficiency of computer-based systems for education, information, commerce, etc., strongly depends on the user-friendliness of the corresponding interfaces, in particular, on the location and size of the appropriate icons. When deciding the location and size of different icons on a computer screen, designers use the Fitts's Law [41, 42]. This law describe how the average time T of accessing an icon depends on the distance d from the center of the screen to the icon and on the linear size w of this icon: $T = a + b \cdot \ln\left(\frac{d}{w}\right)$, for some constants a and b.

How Fitts's Law is used in interface design. The use of Fitts's Law started with the very first mouse-accessible interfaces; see, e.g., [43]. It is based on the following idea.

Each icon corresponds to a specific task or group of tasks. Some tasks are more frequent, some are rarer: for example, editing is a frequent task, while logging off is a rarer task. For each task, we can empirically determine the frequency f_i with which this task is performed. We can therefore gauge the user-friendliness of the interface by the average time

$$\sum_i f_i \cdot T_i = \sum_i f_i \cdot \left(a + b \cdot \ln\left(\frac{d_i}{w_i}\right)\right)$$

needed to access the required icon. Out of several possible interfaces, we select the one for which this average time is the smallest.

Fitts's Law: qualitative aspects. From the qualitative viewpoint, the Fitts's Law says that T decreases when d decreases and/or w increases. In other words:

- the closer the icon to the center, the easier it is to find this icon, and
- the larger the icon size, the easier it is to find it.

From this viewpoint, Fitts's Law is simply common sense.

Quantitative aspects of the Fitts's Law need explanation. That the time T should monotonically depend on the distance d and on the size w is clear, but there are many different monotonic functions. The fact that overwhelming majority of experimentally results is in very good accordance with one type of monotonic dependence—the logarithmic law—needs explanation.

Current explanation of Fitts's Law. A current explanation of Fitts's Law [44] is is based on the fact that our motions are not perfect. For simplicity, this explanation assumes that each movement aiming at reaching an object at distance d actually only follows a slightly smaller distance $(1 - \varepsilon) \cdot d$, for some accuracy $\varepsilon < 1$. Thus, after the original movement, we are still a distance $(\varepsilon \cdot d$ away from the desired object. We therefore need the next movement to reach this object.

This second movement brings us to the distance $\varepsilon \cdot (\varepsilon \cdot d) = \varepsilon^2 \cdot d$ to the target. In general, after k movements, we are at a distance $\varepsilon^k \cdot d$ from the target. If we aim at the center of an icon, then we reach a point within the icon when this distance is smaller than or equal to the icon's half-size $\frac{w}{2}$, i.e., when $\varepsilon^k \cdot d \leq \frac{w}{2}$. From the condition that

$\varepsilon^k \cdot d \approx \frac{w}{2}$, we can determine the number of iterations k as $k \approx \frac{1}{\ln(\varepsilon)} \cdot \ln\left(\frac{2d}{w}\right)$. One can easily check that we thus get $k \approx a + b \cdot \ln\left(\frac{d}{w}\right)$, where $a = \frac{\ln(2)}{\ln(\varepsilon)}$ and $b = \frac{1}{|\ln(\varepsilon)|}$.

The overall time needed to reach the icon consists of the time of the smooth motions and the time needed to switch from one motion to another. Usually, the switch time is much larger. So, in first approximation, we can simply ignore the time of the smooth movements and conclude that the time T is proportional to the number of switches k. Thus, we arrive at the Fitts's formula.

This explanation has some gaps. As noted, e.g., in [45], the above explanation is not perfect, it has two gaps. The first gap is not critical: the above derivation assumes that for the same distance d, the motor error is always the same, while in reality, in repeated experiments, we may get different values of the motor error. This gap is not critical, because the above derivation will not change is we take into account that after the first iteration, the distance to the target is only *approximately* equal to $\varepsilon \cdot d$.

The second gap is more serious. The above derivation is based on the assumption that if we want to move to a distance d, then the accuracy with which we can perform this movement is equal to $\varepsilon \cdot d$. In other words, this derivation is based on the assumption that the *relative* accuracy $\frac{\varepsilon \cdot d}{d}$ is the same for all the distances. If the relative accuracy depends on the distance d, i.e., if the accuracy is equal to $\varepsilon(d) \cdot d$ for some function $\varepsilon(d) \neq \text{const}$, then, instead of the Fitts's Law, we would get a different formula.

What we do. To come up with a more convincing explanation of the Fitts's Law, we therefore need to explain why the relative accuracy does not depend on the distance. This is what we do in this section.

Our explanation of Fitts's Law. Let us assume that the cursor (controlled, e.g., by a mouse) is currently located at the center C of the screen, and we want to move it to the location of the icon I. The shortest way from one point to another is a straight line, so naturally, we start a straight line in the direction of the icon. To be more precise, we select an angle leading us to the icon, and we follow a straight line in the direction of this angle.

If we could set up the angle exactly, we would then follow the straight line to the desired icon and reach this icon in one movement. In practice, of course, there is a motor error; we cannot set the angle of our movement exactly, we can only set up this angle with some accuracy θ. Because of this accuracy, the straight line that we actually follow is at an angle of order θ from the line connecting the center of the screen with the target icon.

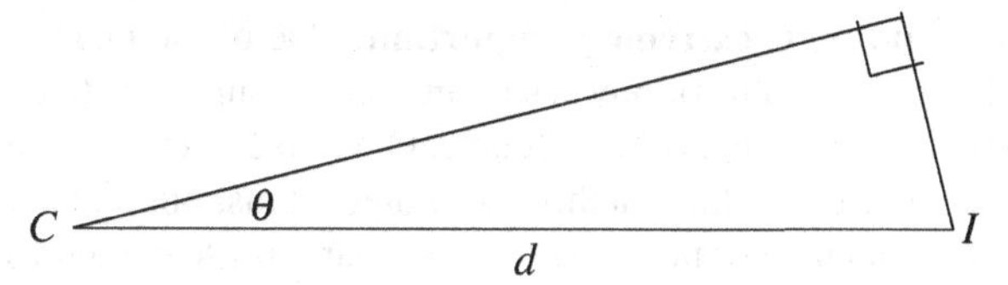

As a result of this motion inaccuracy, we do not reach the desired point I, the closest we get to I is at a distance $\approx d \cdot \sin(\theta)$. As a result of a movement, we get from the location at a distance d from the target point I to a new location whose distance to I is approximately equal to $\varepsilon \cdot d$, where $\varepsilon \stackrel{\text{def}}{=} \sin(\theta)$.

To reach the desired location I, starting from this new point, we again aim at I. As a result, we get from the point at a distance $\approx \varepsilon \cdot d$ to I to a new point whose distance from I is approximately equal to $\varepsilon \cdot (\varepsilon \cdot d) = \varepsilon^2 \cdot d$. After k iterations, we reach a point at a distance $\approx \varepsilon^k \cdot d$ to the target point I. We reach the icon if this distance does not exceed the icon's half-width $\frac{w}{2}$, i.e., when $\varepsilon^k \cdot d \approx \frac{w}{2}$.

As we have mentioned, the resulting number of iterations is $k \approx a + b \cdot \ln\left(\frac{w}{d}\right)$. Under a natural assumption that the average time T needed to reach an icon is proportional to this number of iterations, we get the desired Fitts's Law.

Comment. It is worth mentioning that a similar geometric argument describes how the number of corrections needed for inter-stellar travel depends on the travel distance d; see, e.g., [46].

4.5 Analyzing the Effect of Feedback

The efficiency of presenting information to students depends on how we present the material and on whether we use the appropriate feedback—and how exactly we use it. In the previous two sections, we analyzed different ways of presenting the material. In this section, we analyze the effect of feedback.

A recent study [47] published in the *Notices of American Mathematical Society* showed among many factors which could potentially affect the students' success, only one factor determines the success of a technique: the presence of immediate feedback. On average, students who receive immediate feedback learn twice faster than students who are taught in a more traditional way, with a serious feedback only once or twice a semester (after a test).

The very fact that immediate feedback is helpful is not surprising: it helps the student clear misconceptions and avoid the wrong paths. However, the fact that different techniques involving feedback lead to practically the same learning speed-up is intriguing. To explain this speed-up, we provide a simplified first-order description of a learning process in simple geometric terms. We show that already in this first approximation, the geometric description leads to the observed two-fold speed-up in learning.

The results from this section were first published in [48, 49].

Student understanding is extremely important. One of the main objectives of a course—whether it is calculus or physics or any other course—is to enable students to understand the main concepts of this course. Of course, it is also desirable that the students learn the corresponding methods and algorithms, but understanding is the primary goal. If a student does not remember by heart how to compute the derivative of a product, he or she can look up the formula on the web or even derive the formula – and so, most probably, this student will succeed in the following classes which depend on the use of derivatives. However, if a student does not have a good understanding of what is a derivative, then even if this student remembers some formulas, the student will probably not be able to decide which formula to apply in what situation.

How to gauge student understanding. To properly gauge student's understanding, several disciplines have developed *concept inventories*, sets of important basic concepts and questions testing the students' understanding of these concepts. The first such concept inventory was developed in physics, to gauge the students' understanding of the basic concepts of Newtonian mechanics such as the concept of force; the corresponding Force Concept Inventory (FCI) is described in [50–54]. A similar Calculus Concept Inventory (CCI) is described in [55, 56].

A student's degree of understanding is measured by the percentage of the questions that are answered correctly. The class's degree of understanding is measured by averaging the students' degrees. An ideal situation is when everyone has a perfect 100% understanding; in this case, the average score is 100%. In practice, the average score is smaller than 100%.

How to compare different teaching techniques. To gauge how successful is a given teaching technique, we can measure the average score μ_0 before the class and the average score μ_f after the class. A perfect class is when the whole difference $100 - \mu_0$ disappeared, i.e., the students' average score went from μ_0 to $\mu_f = 100$. In practice, of course, the students' gain $\mu_f - \mu_0$ is somewhat smaller than the ideal gain $100 - \mu_0$. It is reasonable to measure the success of a teaching method by describing which portion of the ideal gain is covered, i.e., by the ratio

$$g \stackrel{\text{def}}{=} \frac{\mu_f - \mu_0}{100 - \mu_0}.$$

Empirical results. It turns out that for different teaching methods, the normalized gain g does not depend on the initial level μ_0, does not depend on the textbook used or on the teacher. Only one factor determines the value g: the absence or presence of immediate feedback.

In the traditionally taught classes, where the students get their major feedback only after their first midterm exam, the normalized gain g is consistently smaller than in the classes where the students got immediate feedback during every class period.

Specifically, for traditionally taught classes, the average value of the gain is $g \approx 0.23$, while for the classes with an immediate feedback, the average value of the gain is $g \approx 0.48$; see, e.g., [47, 50].

In other words, students who receive immediate feedback, on average, learn twice faster than students who are taught by traditional methods.

Natural question. The consistent appearance of the doubling of the rate of learning seems to indicate that there is a fundamental reason behind this empirical result.

What we do in this section. In this section, we provide a possible geometric explanation for the above empirical result.

Why geometry. Learning means that the student—who did not originally know the material—becomes knowledgeable of this material. To check how well a student learned, we can apply different tests. Based on the results of these tests, we can determine the current state of the student knowledge. In other words, at any given moment of time, the state of the student's knowledge can be characterized by several numbers $(x_1, \ldots, x_n)$—the student's scores on different parts of the test.

Each such state can be naturally represented as a point in the n-dimensional space—namely, a point with coordinates $x_1, \ldots, x_n$. In the starting state S, the student does not know the material; the desired state D describes the situation when a student has the desired knowledge. When a student learns, the student's state of knowledge changes continuously forming a (continuous) trajectory γ which starts at the starting state S and ends up at the desired state D.

First simplifying assumption: all students learn at the same rate. Some students learn faster, others learn slower. The above empirical fact is not about their individual learning rates, it is about the *average* rates of student learning, averaged over all kinds of students. From this viewpoint, it makes sense to simplify the complex actual situation—in which different students have different learning rates—with a simplified model, in which all the students have the same average learning rate.

Let us give an example of why such a replacement makes sense when we only consider averages:

- if we are want to study the difference between people's appetites, it makes sense to keep their differing heights intact;
- however, if we are planning to serve a meal to a large group of people, it makes perfect sense, when ordering food ingredients, to ignore the individual differences and assume that everyone has an average appetite.

In geometric terms, the rate of learning corresponds to the rate with which the student's state changes, i.e., corresponds to how far the student's state of knowledge changes in a given period of time. In these terms, the assumption that all the students have the same learning rate means that the states corresponding to different students change with the same rate. In other words, in this geometric model, the time that it takes for a student to get from the initial state S to the desired state D is proportional to the total length of the corresponding curve γ.

In these terms, to explain the fact that students who receive instant interaction learn twice faster means that on average, we need to show that their learning trajectories are, on average, twice shorter.

Second simplifying assumption: the shape of the learning trajectories. In the beginning, a student may be eager to study, but often, he/she is not sure which direction to go. A student usually has misconceptions about physics and/or calculus, misconception that may lead the student in a wrong direction. We can describe this by assuming that when a student starts at the starting point S, he/she moves in a random direction.

In situations when the student deviated from the direction towards the desired state D, a feedback enables the student understand that he/she is going in the wrong direction. After the feedback, the student corrects his/her trajectory.

In the case of immediate feedback, this correction comes right away, so, in effect, the student immediately starts following the right direction. In other words, in learning with immediate feedback, the student's learning trajectory is a straight line which goes directly from S to D.

In the traditional learning, feedback comes only with midterm exams. Usually, there are two midterm exams, and they are scheduled in such a way that between themselves, they cover all the material studied in the course, and each covers approximately the same amount of material. Thus, the first midterm exam usually covers half of the material. In geometric terms, it means that this exam is given once the student covered half of the distance between S and D. This exam checks whether the student has correctly reached the midpoint $M \stackrel{\text{def}}{=} \frac{S+D}{2}$ between S and D. Once the student has covered the half-distance $d/2$ in the originally selected direction, the results of the first midterm exam provides a necessary correction, and the student starts going straight towards the correct midpoint M. After that, the same process starts again: the student goes for $d/2$ in the random direction, and then comes back to D.

Resulting geometric description of learning with and without immediate feedback. In learning with immediate feedback, a student follows a straight line from S to D. The length of the corresponding trajectory is equal to the distance $d \stackrel{\text{def}}{=} \rho(S, D)$ between the states S and D.

In learning without immediate feedback, a student first follows a straight line of length $d/2$ which goes in a random direction, then goes straight to the midpoint M, then again follows a straight line of length $d/2$ in a random direction, and finally takes a straight like to D.

Third simplifying assumption: the state space is 1-D. While in general, we can think of different numerical characteristics describing different aspects of student knowledge, in practice, we are pretty comfortable using a single number—usually, an overall grade for the course—to characterize the student's state of knowledge. It is therefore reasonable to make one more simplifying assumption: that the state of a student is characterized by only one parameter x_1.

Let us compare the lengths of the corresponding trajectories. Under our simplifying assumption, the learning time is proportional to the length of the corresponding trajectory. Thus, to compare the learning rates, we need to compare the lengths of the corresponding trajectories.

In case of immediate feedback, the learning trajectory has length d. So, to make a comparison, we must estimate the length of a trajectory corresponding to the traditional learning.

This trajectory consists of two similar parts: the part connecting S and M and the part connecting M and D. Hence, to estimate the total average length, it is sufficient to estimate the average length from S to M and then multiply the result by two.

Analysis of the model. In case of immediate feedback, the learning trajectory has length d.

In the case of traditional learning, under the 1-D assumption, a student initially goes either in the correct direction or in the opposite (wrong) direction; the idea that the direction is chosen randomly can be naturally formalized as an assumption that both directions occur with equal probability 1/2.

If the student's trajectory initially moves in the correct direction, then after traveling the distance $d/2$, the state gets exactly into the desired midpoint D; so, the overall length of the S-to-M of the trajectory is exactly $d/2$.

If the student's trajectory initially goes in the wrong direction, then the student ends up at a point at the same distance $d/2$ from S but on the wrong side of S. Getting back to M then means first going back to S and then going from S to M. The overall length of this trajectory is thus $3d/2$.

With probability 1/2, the length is $d/2$, with probability 1/2, the length is $3d/2$. So, the average length of the S-to-M part of the learning trajectory is equal to

$$\frac{1}{2} \cdot \frac{d}{2} + \frac{1}{2} \cdot \frac{3d}{2} = d.$$

The average length of the whole trajectory is double that, i.e., $2d$—twice larger than the length corresponding to immediate feedback.

Since we assumed that the learning time is proportional to the length of the learning trajectory, we can thus draw the following conclusion.

Summary of the results. In this 1-D model, a student following an instant feedback trajectory reaches the desired state, on average twice faster than a student following the traditional-learning trajectory. This is exactly what we wanted to explain.

References

1. L.O. Lerma, D. Pennington, V. Kreinovich, On early stages of idea propagation, the number of adopters grows as $n(t) \approx c \cdot t^a$: theoretical explanation of the empirical observation. J. Uncertain. Syst. **8**(3), 180–185 (2014)
2. G. Aletti, A.K. Naimzada, G. Naldi, Mathematics and physics applications in sociodynamics simulation: the case of opinion formation and diffusion, in *Mathematical Modeling of Collective Behavior in Socio-Economic and Life Sciences*, ed. by G. Naldi, L. Pareschi, G. Toscani (Springer, Dordrecht, 2010), pp. 203–221

3. L.M.A. Bettencourt, A. Cintron-Arias, D.I. Kaiser, C. Castillo-Chavez, The power of a good idea: quantitative modeling of the spread of ideas from epidemiological models. Physica A **364**, 513–536 (2006)
4. L.M.A. Bettencourt, D.I. Kaiser, J. Kaur, C. Castillo-Chavez, D.E. Wojick, Population modeling of the emergence and development of scientific fields. Scientometrics **75**(3), 495–518 (2008)
5. P.A. Geroski, Models of technology diffusion. Res. Policy **29**, 603–625 (2000)
6. M. Gomez-Rodriguez, J. Leskovec, A. Krause, Inferring networks of diffusion and influence, in *Proceedings of the 16th ACM SIGKDD International Conference on Knowledge Discovery and Data Mining KDD'10*, Washington, DC, July 25–28, 2010
7. M. Ingawale, Understanding the Wikipedia phenomenon: a case for agent based modeling, in *Proceedings of the ACM Ph.D. Workshop on Information and Knowledge Management PIKM'08*, Napa Valley, California, October 30, 2008, pp. 81–84
8. P. Morone, R. Taylor, *Knowledge Diffusion and Innovation: Modeling Complex Entrepreneurial Behaviors* (Edward Elgar Publ, Cheltenham, UK, 2010)
9. G. Rossman, *Climbing the Charts: What Radio Airplay Tells Us about the Diffusion of Information* (Princeton University Press, Princeton, New Jersey, 2012)
10. C.M. Tsai, The knowledge diffusion model associated with innovative knowledge. Expert Syst. Appl. **36**, 11323–11331 (2009)
11. J. Xu, M. Quaddus, Examining a model of knowledge management systems adoption and diffusion: a partial Least Square approach. Knowl.-Based Syst. **27**, 18–28 (2012)
12. P. Yan, X. Yang, To trade or to teach: modeling tacit knowledge diffusion in complex social networks, in *Proceedings of the Second International Conference on Future Information Technology and Management Engineering FITME'2009*, Sanya, China, December 13–14, 2009, pp. 151–154
13. J. Jee, L.C. Klippel, M.S. Hossain, N. Ramakrishnan, B. Mishra, Discovering the ebb and flow of ideas from text corpora, in *IEEE Computer*, February 2012, pp. 77–81
14. B.K. Chakrabarti, A. Chakraborti, A. Chatterjee, *Econophysics and Sociophysics: Trends and Perspectives* (Wiley-VCH, Berlin, 2006)
15. S.I. Resnick, *Heavy-Tail Phenomena: Probabilistic and Statistical Modeling* (Springer-Verlag, New York, 2007)
16. B.B. Mandelbrot, *The Fractal Geometry of Nature* (W. H. Freeman Publ, San Francisco, California, 1983)
17. D.J. Sheskin, *Handbook of Parametric and Nonparametric Statistical Procedures* (Chapman & Hall/CRC, Boca Raton, Florida, 2011)
18. O. Lerma, L. Valera, D. Pennington, V. Kreinovich, How knowledge propagates? a fractal model justified on the example of the Out of Eden Walk, in *Abstracts of the 12th International Sun Conference on Teaching and Learning*, El Paso, Texas, March 5–6, 2015
19. O. Lerma, L. Valera, D. Pennington, V. Kreinovich, How knowledge propagates? a fractal model justified on the example of the Out of Eden Walk, in *Abstracts of the NMSU/UTEP Workshop on Computer Science, Mathematics, and Computational Science*, Las Cruces, New Mexico, April 11, 2015
20. Out of Eden Walk, http://www.outofedenwalk.com/
21. M. Le Masurier, What is Slow Journalism?, in *Journalism Practice*, 2014
22. Pulitzer Centre, *Out of Eden: Paul Salopeks Walk from Ethiopia to Patagonia*, http://pulitzercenter.org/projects/out-of-eden
23. H.S. Tavangar, B. Mladic-Morales, *The Global Education Toolkit for Elementary Learners* (Corwin, Thousand Oaks, California, 2014)
24. P.J. Huber, E.M. Ronchetti, *Robust Statistics* (Wiley, Hoboken, New Jersey, 2009)
25. O. Lerma, O. Kosheleva, S. Shahbazova, V. Kreinovich, Towards designing optimal individualized placement tests, in *Abstracts of the 11th International Sun Conference on Teaching and Learning*, El Paso, Texas, March 5–7, 2014, p. 12
26. O. Lerma, O. Kosheleva, S. Shahbazova, V. Kreinovich, *Towards Designing Optimal Individualized Placement Tests*, University of Texas at El Paso, Department of Computer Science, Technical Report UTEP-CS-13-75b, 2013

27. T.H. Cormen, C.E. Leiserson, R.L. Rivest, C. Stein, *Introduction to Algorithms* (MIT Press, Cambridge, MA, 2009)
28. H.T. Nguyen, V. Kreinovich, E. Kamoroff, Asymmetric information measures: how to extract knowledge from an expert so that the expert's effort is minimal. Int. J. Autom. Control (IJAAC) **2**(2/3), 153–177 (2008)
29. R.A. Trejo, J. Galloway, C. Sachar, V. Kreinovich, C. Baral, L.C. Tuan, From planning to searching for the shortest plan: an optimal transition. Int. J. Uncertai. Fuzziness Knowl.-Based Syst. (IJUFKS) **9**(6), 827–838 (2001)
30. O. Lerma, O. Kosheleva, V. Kreinovich, Interleaving enhances learning: a possible geometric explanation. Geombinatorics **24**(3), 135–139 (2015)
31. K.G. Hall, D.A. Dominguez, R. Cavazos, Contextual interference effects with skilled basketball players. Perceptual and Motor Skills **78**, 835–841 (1994)
32. D.K. Landin, E.P. Hebert, M. Fairweather, The effects of variable practice on the performance of a basketball skill. Res. Q. Exercise Sports **64**, 232–236 (1993)
33. N. Kornell, R.A. Bjork, Leaning concepts and categories: is spacing the 'enemy of induction'? Psychol. Sci. **19**, 585–592 (2008)
34. K. Le Blanc, D. Simon, Mixed practice enhances retention and JOL accuracy for mathematical skills, in *Proceedings of the 49th Annual Meeting of the Psychonomic Society*, Chicago, Illinois, November 2008
35. D. Rohrer, K. Taylor, The shuffling of mathematics practice problems boosts learning. Instruct. Sci. **35**, 481–498 (2007)
36. K. Taylor, D. Rohrer, The effects of interleaved practice. Appl. Cognit. Psychol. **24**, 837–848 (2010)
37. M.S. Birnbaum, N. Kornell, E.L. Bjork, R.A. Bjork, Why interleaving enhances inductive learning: the roles of discrimination and retrieval. Memory Cognit. **41**, 392–402 (2013)
38. O. Lerma, V. Kreinovich, Student autonomy improves learning: a theoretical justification of the empirical results. J. Uncertain. Syst. **10** (2016)
39. J. Reeve, H. Jang, What teachers say and do to support students' autonomy during a learning activity. J. Educ. Psychol. **98**(1), 209–218 (2006)
40. E.T. Jaynes, G.L. Bretthorst, *Probability Theory: The Logic of Science* (Cambridge University Press, Cambridge, UK, 2003)
41. P.M. Fitts, The information capacity of the human motor system in controlling the amplitude of movement. J. Exp. Psychol. **47**(6), 381–391 (1954)
42. P.M. Fitts, J.R. Peterson, Information capacity of discrete motor responses. J. Exp. Psychol. **67**(2), 103–112 (1964)
43. S.K. Card, W.K. English, B.J. Burr, Evaluation of mouse, rate-controlled isometric joystick, step keys, and text keys for text selection on a CRT. Ergonomics **21**(8), 601–613 (1978)
44. S.K. Card, T.P. Moran, A. Newell, *The Psychology of Human-Computer Interaction* (CRC Press, Boca Raton, Florida, 1986)
45. D.E. Meyer, J.E.K. Smith, S. Kornblum, R.A. Abrams, C.E. Wright, Speed-accuracy tradeoffs in aimed movements: Toward a theory of rapid voluntary action, in *Attention and Performance*, ed. by M. Jeannerod (Lawrence Erlbaum Publ, Hillsdale, New Jersey, 1990), pp. 173–226
46. O. Kosheleva, V. Kreinovich, R. Labiaga, Inter-stellar travel and simple geometrical combinatorics: how many trajectory corrections are necessary? Geombinatorics **4**(4), 129–136 (1994)
47. J. Epstein, The calculus concept inventory - measurement of the effect of teaching methodology in mathematics. Notices Am. Math. Soc. **60**(8), 1018–1026 (2013)
48. O. Lerma, O. Kosheleva, V. Kreinovich, A simple geometric model provides a quantitative explanation of the advantages of immediate feedback in student learning, in *Abstracts of the 11th International Sun Conference on Teaching and Learning*, El Paso, Texas, March 5–7, 2014, p. 11
49. O. Lerma, O. Kosheleva, V. Kreinovich, A simple geometric model provides a possible quantitative explanation of the advantages of immediate feedback in student learning. Geombinatorics **25**(1), 22–29 (2015)

50. R.R. Hake, Interactive engagement versus traditional methods: a six-thousand student survey of mechanics test data for physics courses. Am. J. Phys. **66**, 64–74 (1998)
51. I. Halloun, D. Hestenes, The initial knowledge state of college physics students. Am. J. Phys. **53**, 1043–1055 (1985)
52. I. Halloun, D. Hestenes, Common sense concepts about motion. Am. J. Phys. **53**, 1056–1065 (1985)
53. D. Hestenes, I. Halloun, Interpreting the FCI. Phys. Teacher **33**, 502–506 (1995)
54. D. Hestenes, M. Wells, G. Swackhamer, Force concept inventory. Phys. Teacher **30**, 141–158 (1992)
55. J. Epstein, Development and validation of the Calculus Concept Inventory, in *Proceedings of the Ninth International Conference on Mathematics Education in a Global Community*, ed. by D.K. Pugalee, A. Rogerson, A. Schinck (North Carolina, September, Charlotte, 2007), pp. 7–12
56. J. Epstein, The calculus concept inventory—new data, correlation with teaching methodology, in *Abstracts of the Joint Mathematics Meeting: 115th Annual Meeting of the American Mathematical Society (AMS) and the 92nd Meeting of the Mathematical Association of America (MAA)*, Washington, DC, January 5–8, 2009

Chapter 5
Knowledge Use

Abstract How can we use the acquired knowledge? In many practical situations, we have a well-defined problem, with a clear well-formulated objective. Such problems are typical in engineering: we want a bridge which can withstand a given load, we want a car with a given fuel efficiency, etc. There exist many techniques for solving such well-defined optimization problems. However, in many practical situations, we only have partial and/or subjective information about the situation and about our objectives. In such situations, we need to make decisions under uncertainty. This aspect of knowledge use is what we analyze in this chapter. The ultimate goal of knowledge use is to help the users. To do this, we need to get a good understanding of the corresponding processes. Gaining such an understanding is the main objective of *science*, when we use the observed data to find the dependencies between different quantities. Once these dependencies have been discovered, we can apply this knowledge to help the users: we can find out how to better *control* the existing systems, we can find out how to better *design* the new systems, and we can find out how to better *maintain* the systems. In all these engineering tasks, we are interested in decision making under uncertainty, in particular, in taking imprecise expert knowledge into account. In this chapter, we provide examples of applications to science and to all three aspects of engineering. Specifically, in Sect. 5.1, we consider an example of an application to science, in Sect. 5.2, we consider an example of an application to control, in Sect. 5.3, we deal with an application to design, and in Sect. 5.4, we consider an application to maintenance.

5.1 Use of Knowledge in Science: A Case Study

In this section, as a case study, we consider an important problem related to global warming. This is an area in which there is clearly still a lot of uncertainty about the relevant physical processes. Specifically, in this section, we deal with the following

L.O. Lerma and V. Kreinovich, *Towards Analytical Techniques for Optimizing Knowledge Acquisition, Processing, Propagation, and Use in Cyberinfrastructure and Big Data*, Studies in Big Data 29, DOI 10.1007/978-3-319-61349-9_5

problem: while global warming is a statistically confirmed long-term phenomenon, its most visible consequence is not the warming itself but, somewhat surprisingly, the increased climate variability. In this section, we use general system theory ideas to explain why increased climate variability is more visible than the global warming itself.

The results from this section first appeared in [1].

What is global warming. The term "global warming" usually refers to the fact that there is a statistically significant long-term increase in the average temperature; see, e.g., [2–5].

Somewhat surprisingly, what we mainly observe is not global warning itself, but rather related climate variability. Researchers have analyzed the expected future consequences of global warming: increase in temperature, melting of glaciers, raising sea level, etc. A natural hypothesis was that at present, when the effects of global warming are just starting, we would see the same effects, but at a smaller magnitude. This turned out not to be the case.

Some places do have the warmest summers and the warmest winters in record. However, other places have the coldest summers and the coldest winters on record.

What we do observe in all these cases is not so much the direct effects of the global warming itself, but rather an increased climate variability, an increase not so much in the *average* temperatures but rather in the *variance* of the temperature: in both unusually warm days and unusually cold days, what we observe is unusually high deviations from the average.

This phenomenon is sometimes called *climate change*, but a more proper description should be *increased climate variability* [2–5].

Why is increased climate variability more visible than global warming? A natural question. A natural question is: why is increased climate variability more visible than global warming—which is supposedly the reason for this increased variability?

A usual answer to this question—and its limitations. A usual answer to the above question is that the increased climate variability is what computer models predict. However, the existing models of climate change are still very crude. Their quantitative predictions are usually very approximate and often unreliable, even on the qualitative scale: for example, none of these models explains the fact that the growth in the average temperature has drastically slowed down in the last two decades [2–5].

It is therefore desirable to supplement the usual computer-model-based answer to the above question by more reliable explanations.

What we do in this section. In this section, we show that, on the qualitative level, the fact that the increase climate variability is more observable than the global warming can be explained in general system-theoretic terms.

Towards a simplified model: first approximation. For simplicity, let us consider the simplest possible model, in which the state of the Earth is described by a single parameter x. In our case, x can be an average Earth temperature or the temperature at a certain location.

We want to describe how the value x of this parameter changes with time. In other words, we want to describe the derivative $\frac{dx}{dt}$.

There are external forces affecting the dynamics. So, in the first approximation, we can say that $\frac{dx}{dt} = u(t)$, where $u(t)$ describes these external forces.

We know that, on average, these forces lead to a global warming, i.e., to the increase in the value of the parameter $x(t)$. In terms of our equation, this means that the average value of $u(t)$ is positive. Let us denote this average value by u_0, and the random deviation from this average by $r(t) \stackrel{\text{def}}{=} u(t) - u_0$, then $u(t) = u_0 + r(t)$.

For simplicity, we will assume that the random values $r(t)$ corresponding to different moments t are independent and identically distributed, with some standard deviation σ_0.

Towards the second approximation. The above simplified equation does not take into account the fact that most natural systems—including the system corresponding to climate—are somewhat resistant to change: if a system is not resistant to change, it would not have persisted in the presence of numerous external forces.

Resistance to change means that when a system deviates from its stable value x_0, forces appear that try to bring this system back to this stable value. From the mathematical viewpoint, this phenomenon is easier to describe if instead of the original variable x, we consider the difference $y \stackrel{\text{def}}{=} x - x_0$. In terms of this difference, when $y > 0$, we have a force that decreases y, and when $y < 0$, we have a force that increases y. When $y = 0$, i.e., when $x = x_0$, the system remains in the stable state, so there are no forces.

In precise terms, in the absence of external forces, the system's dynamics is described by an equation $\frac{dy}{dt} = f(y)$, where $f(0) = 0$, $f(y) < 0$ for $y > 0$, and $f(y) > 0$ for $y < 0$. Since the system is stable, deviations y from the stable state are relatively small, so we can expand $f(y)$ in Taylor series in y and retain only the first few terms in this expansion. In general, we have $f(y) = f_0 + f_1 \cdot y + \ldots$ The condition $f(0) = 0$ implies that $f_0 = 0$, so $f(y) = f_1 \cdot y + \ldots$ The condition that $f(y) < 0$ for $y > 0$ implies that $f_1 < 0$, i.e., that $f_1 = -k$ for some $k > 0$. Thus, by keeping only the leading term in the Taylor expansion, we get $f(y) = -k \cdot y$.

Thus, we arrive at the following equation.

Resulting equation.

$$\frac{dy}{dt} = -k \cdot y + u_0 + r(t). \tag{5.1.1}$$

Discussion. Due to the linear character of the Eq. (5.1.1), each solution of this equation can be represented as a sum $y(t) = y_s(t) + y_r(t)$ of the solutions $y_s(t)$ and $y_r(t)$

corresponding to the systematic (average) part u_0 of the outside force and and to the random part $r(t)$:

$$\frac{dy_s}{dt} = -k \cdot y_s + u_0; \tag{5.1.2}$$

$$\frac{dy_r}{dt} = -k \cdot y_r + r(t). \tag{5.1.3}$$

Here, the systematic component $y_s(t)$ describes the systematic change in temperature (global warming), while the random component $y_r(t)$ describes the random change in temperature, i.e., the increased climate variability.

An empirical fact that needs to be explained. We need to explain that, in spite of the fact that eventually, we will see the effects of the global warming itself, at present, the climate variability becomes more visible than the global warming itself. In other words, at present, the relative role $y_r(t)/y_s(t)$ of climate variability is much higher than it will be in the future, when the global warming may become significant.

How to describe this empirical fact in precise terms? The change in y is determined by two factors: the external force $u(t)$ and the parameter k that describes how resistant is our system to this force (the larger k, the large resistance to the change).

While some part of global warming may be caused by the variations in Solar radiation, most climatologists agree that the prevailing part of the long-term global warming is caused by local processes—such as the greenhouse effect—that lower the system's natural resistance to changes. (What causes numerous debates is which proportion of the global warming is caused by human activities.)

Since the decrease in resistance is the major contribution to the observed phenomena, in the first approximation, we will consider only this decrease. In other words, we will assume that the forces remain the same, but the parameter k decreases with time.

In these terms, the observed phenomenon is that at present, when the resistance value k is still reasonably high, the ratio $y_r(t)/y_s(t)$ is much larger than it will be in the future, when the resistance k will decrease. In other words, what we need to explain is that this ratio decreases when the value k decreases.

When computing this ratio, we need to take into account that while the systematic component $y_s(t)$ is deterministic, the random component $y_r(t)$ is a random process, its values change wildly. To gauge the size of this random component, i.e., to gauge how far the random variable $y_r(t)$ deviates from 0, it is reasonable to use standard deviation $\sigma(t)$ of this random variable.

Thus, we arrive at the following formulation.

Resulting formulation of the problem. We fix values u_0 and σ_0. Then, for each k, we can form the solutions $y_s(t)$ and $y_r(t)$ of the differential Eqs. (5.1.2) and ((5.1.3) corresponding to $y_s(0) = 0$ and $y_r(0) = 0$, where $r(t)$ is a family of independent identically distributed random variables with 0 mean and standard deviation σ_0. Since $r(t)$ is random, the solution $y_r(t)$ is also random, so for each moment t, we can define the standard deviation $\sigma(t)$ of this solution.

We want to prove that for every moment t, for sufficiently large $k > 0$, when k decreases, then the ratio $\sigma(t)/y_s(t)$ also decreases.

Solving the equation for the systematic deviation $y_s(t)$. If we move all the terms containing the unknown $y_s(t)$ to the left-hand side, we get

$$\frac{dy_s(t)}{dt} + k \cdot y_s(t) = u_0. \tag{5.1.4}$$

For an auxiliary variable $z(t) \stackrel{\text{def}}{=} y_s(t) \cdot \exp(k \cdot t)$, we get

$$\frac{dz(t)}{dt} = \frac{dy_s(t)}{dt} \cdot \exp(k \cdot t) + y_s(t) \cdot \exp(k \cdot t) \cdot k =$$

$$\exp(k \cdot t) \cdot \left(\frac{dy_s(t)}{dt} + k \cdot y_s(t)\right). \tag{5.1.5}$$

Thus, due to (5.1.4), we have

$$\frac{dz(t)}{dt} = u_0 \cdot \exp(k \cdot t). \tag{5.1.6}$$

We know that for $t = 0$, we have $y_s(0) = 0$ and thus, $z(0) = 0$. Thus, the value $z(t)$ can be obtained by integration:

$$z(t) = z(0) + \int_0^t u_0 \cdot \exp(k \cdot s)\, ds = u_0 \cdot \frac{\exp(k \cdot t) - 1}{k}. \tag{5.1.7}$$

Hence, for $y_s(t) = \exp(-k \cdot t) \cdot z(t)$, we get

$$y_s(t) = u_0 \cdot \frac{1 - \exp(-k \cdot t)}{k}. \tag{5.1.8}$$

Solving the equation for the random component $y_r(t)$. For the random component, we similarly get

$$y_r(t) = \exp(-k \cdot t) \cdot \int_0^t r(s) \cdot \exp(k \cdot s)\, ds. \tag{5.1.9}$$

The mean value of each variable $r(s)$ is 0, thus, the mean value $E[y_r(t)]$ of their linear combination $y_r(t)$ is also 0. Hence, the variance

$$\sigma^2(t) = E[(y_r(t) - E[y_r(t)])^2]$$

of the random component $y_r(t)$ is simply equal to the expected value $E[y_r^2(t)]$ of its square.

Due to the formula (5.1.9), we have

$$y_r(t)^2 = \exp(-2k \cdot t) \cdot \left(\int_0^t r(s) \cdot \exp(k \cdot s)\, ds \right) \cdot \left(\int_0^t r(v) \cdot \exp(k \cdot v)\, dv \right) =$$

$$\exp(-2k \cdot t) \cdot \int_0^t ds \int_0^t dv\, r(s) \cdot r(v) \cdot \exp(k \cdot s) \cdot \exp(k \cdot v). \qquad (5.1.10)$$

Since the expected value of a linear combination is equal to the linear combination of expected values, we get

$$\sigma^2(t) = E[y_r(t)^2] =$$

$$\exp(-2k \cdot t) \cdot \int_0^t ds \int_0^t dv\, E[r(s) \cdot r(v)] \cdot \exp(k \cdot s) \cdot \exp(k \cdot v). \qquad (5.1.11)$$

We assumed that the values $r(s)$ corresponding to different moments of time s are independent and identically distributed, with standard deviation σ_0. Thus, for $s \neq v$, we get $E[r(s) \cdot r(v)] = E[r(s)] \cdot E[r(v)] = 0$ and $E[r^2(s)] = \sigma_0^2$. Substituting these expressions into the formula (5.1.11), we conclude that

$$\sigma^2(t) = E[y_r(t)^2] = \exp(-2k \cdot t) \cdot \int_0^t ds\, \sigma_0^2 \cdot \exp(k \cdot s) \cdot \exp(k \cdot s) =$$

$$\exp(-2k \cdot t) \cdot \int_0^t \sigma_0^2 \cdot \exp(2k \cdot s)\, ds. \qquad (5.1.12)$$

This integral can be explicitly integrated, so we get

$$\sigma^2(t) = \sigma_0^2 \exp(-2k \cdot t) \cdot \frac{\exp(2k \cdot t) - 1}{2k} = \sigma_0^2 \cdot \frac{1 - \exp(-2k \cdot t)}{2k}. \qquad (5.1.13)$$

Analyzing the ratio. We are interested in the ratio $\sigma(t)/y_s(t)$ of two positive numbers. The value $\sigma(t)$ is the square root of the expression (5.1.3). To avoid the need to take square roots, we can take into account the fact that for positive numbers, the square function is increasing; thus, the desired ratio increases with the decrease in k if and only if its square

$$S(t) \stackrel{\text{def}}{=} \frac{\sigma^2(t)}{y_s^2(t)} \qquad (5.1.14)$$

increases. Let us thus analyze this new ratio $S(t)$.

Due to the formulas (5.1.8) and (5.1.13), we get

$$S(t) = \frac{\sigma_0^2}{u_0^2} \cdot \frac{(1 - \exp(-2k \cdot t)) \cdot k^2}{2k \cdot (1 - \exp(-k \cdot t))^2}. \qquad (5.1.15)$$

By using a known formula $a^2 - b^2 = (a - b) \cdot (a + b)$, we conclude that

$$1 - \exp(-2k \cdot t) = (1 - \exp(-k \cdot t)) \cdot (1 + \exp(-k \cdot t)). \quad (5.1.16)$$

Substituting the expression (5.1.16) into the formula (5.1.15) and cancelling the terms k and $1 - \exp(-k \cdot t)$ in the numerator and in the denominator, we conclude that

$$S(t) = \frac{\sigma_0^2}{u_0^2} \cdot \frac{(1 + \exp(-k \cdot t)) \cdot k}{2 \cdot (1 - \exp(-k \cdot t))}. \quad (5.1.17)$$

Conclusion. When the value k is reasonably large, we have $\exp(-k \cdot t) \approx 0$, thus,

$$S(t) \approx \frac{\sigma_0^2}{u_0^2} \cdot \frac{k}{2}. \quad (5.1.18)$$

This ratio clearly decreases when k decreases. Thus, eventually, when the Earth's resistance k will decrease, this ratio will also decrease and so, we will start observing mainly the direct effects of global warming (as researchers originally conjectured)—unless, of course, we do something to prevent the negative effects of global warming.

Comment. In our analysis, we made a simplifying assumption that the climate system is determined by a single parameter x (or y). The conclusion, however, remains the same if we consider a more realistic model, in which the climate system is determined by several parameters $y_1, \ldots, y_n$.

Indeed, in this case, in our linear approximation, the dynamics is described by a system of linear differential equations

$$\frac{dy_i}{dt} = -\sum_{j=1}^{n} a_{ij} \cdot y_j(t) + u_i(t). \quad (5.1.19)$$

In the generic case, all eigenvalues λ_k of the matrix a_{ij} are different; in this case, the matrix can be diagonalized: by considering the linear combinations $z_k(t)$ corresponding to the eigenvectors, we get a system with the diagonal matrix a_{ij}, i.e., a system of the type

$$\frac{dz_k}{dt} = -\lambda_k \cdot z_k(t) + u_k(t). \quad (5.1.20)$$

For each of these equations, similar analysis enables us to reach the same conclusion—that the current ratio of the random to systematic effects is much higher than it will be in the future, when the effects of global warming will be larger.

5.2 Use of Knowledge in Control: A Case Study

In this section, as a case study, we consider an important problem (somewhat related to global warming): the problem of taming the tornados. One of the promising ways to tame a tornado is to inject dust into it. Somewhat counter-intuitively, injecting coarse dust only makes the tornado stronger, while injecting fine dust can indeed help in the taming. This difference has been explained by a mathematical analysis of the corresponding equations, but (in contrast to the usual physics practice) this mathematical analysis has not yet been accompanied by a simple qualitative physical explanation. We show that such a simple explanation can be obtained if we analyze the problem of taming tornados from the geometric viewpoint.

How to tame a tornado: an idea. Blades in an empty blender rotate very fast. However, when we place some material into the blender, the blades slow down. The more material we place in a blender, the more the blades slow down. This known phenomenon leads to a natural idea that maybe we can slow down rotations in a tornado if we inject substance into it.

In general, this idea does not work well. An injection of heavy coarse material into a simulated flow has been tried; unfortunately, it turns out that such an injection only makes the flow stronger; see, e.g., [6, 7]. This is in line with how tornados usually evolve: a tornado starts small, and it power increases as it grabs more matter into its vortex.

This somewhat counter-intuitive experimental fact has been confirmed by a detailed theoretical analysis [8–12], according to which a insertion of coarse dust into a laminar flow indeed has a stabilizing effect on this flow.

There is still hope. Interestingly, the same theoretical analysis shows that insertion of *fine* dust into a flow destabilizes this flow. This fact provides a hope that an injection of fine dust can, in principle, tame the tornado [10–12].

Problem: we need a simple physical explanation. The above-mentioned theoretical explanation is too mathematical for the physicists' taste: it is mathematically correct but it lacks a simple qualitative physical explanation which usually accompanies mathematical analysis of physical phenomena.

What we do. In this section, we show that a geometric analysis of the problem provides the desired simple qualitative explanation.

Describing tornados in geometric terms. From the geometric viewpoint, a tornado is a thick rotating volume of gas. Its 2-D projection is a circle.

This rotating mass also moves, but since the rotation is usually much faster than the linear motion, in the first approximation, we can ignore this linear movement and only concentrate on rotation. In this approximation, the tornado is symmetric with respect to rotations around its axis—and this is its only geometric symmetry.

The ideal state of the atmosphere: a geometric description. According to statistical physics (see, e.g., [13, 14]), all the processes, if undisturbed, eventually reach their most stable state—the state characterized by the maximal entropy.

For the atmosphere, this most stable state is (at least locally) a state with no linear or circular movement, a state in which the density depends only on the height. From the viewpoint of 2-D geometry, this state is invariant with respect to rotations, shifts, and scalings (dilations $\mathbf{x} \to \lambda \cdot \mathbf{x}$).

What we want to achieve when injecting dust. In geometric terms, taming a tornado means that we want to change from the state where the only symmetries are rotations to the state with more symmetries. Eventually, this happens—tornados do not last forever. What we want to do is to speed up this change. We want to achieve this speed-up by introducing a disturbance to the original process—e.g., by injecting dust.

How can we achieve that: a geometric idea. From the geometric viewpoint, if the disturbance has only the same symmetries as the original process, then the introduction of this disturbance does not change the symmetries of the process—and thus, does not lead to the desired objective.

In geometric terms, the only promising way to achieve the desired objective is to add disturbances which have *more* geometric symmetries. Hopefully, this will help the tornado process acquire more symmetries—and thus, speed up the transition to a more symmetric (and hence, less catastrophic) behavior.

The more symmetric the disturbance, the more promising is its influence on the tornado. Let us show that this explains the different effects of coarse and fine dust on tornado taming.

Case of coarse dust. Due to their friction, dust particles becomes smooth. In terms of a 2-D projection, a particle of coarse dust is a small circle.

Its symmetry group is the group of all rotations—exactly the same symmetries group as of the tornado itself. This explains why insertion of the coarse dust does not tame the tornado.

Case of fine dust. By definition, a fine dust consists of particles whose width can be ignored. In geometric terms, such a particle is represented by a point. In terms of the desired 2-D symmetries (rotations, shifts, and dilations), this point is invariant not only with respect to rotations, but also with respect to scalings.

In other words, fine dust has additional symmetries. In view of the above, this explains why injection of the fine dust can indeed destabilize the tornado process.

5.3 Use of Knowledge in Design: A Case Study

In design, it is important to also take into account subjective user preferences. This subjective aspect of decision making is known as *Kansei engineering*. This aspect is what we analyze in this section.

The results from this section first appeared in [15].

Need for Kansei Engineering. Traditional engineering deals with objective characteristics of a design: There may be several different designs with the given ranges on characteristics, e.g., we may have different car designs within the given price range, efficiency range, size restrictions, etc. Different people make different choices between these designs based on their subjective preferences.

This is how people select cars, this is how people select chairs, etc. Engineering that takes such subjective preference into account is known as *Kansei Engineering*; see, e.g., [16–20].

Need to select designs. Different people have different preferences. Thus, to satisfy customers, we must produce several different designs: a car company produces cars of several different designs, a furniture company produces chairs of several different designs, etc.

The creation of each new design is often very expensive and time-consuming. As a result, the number of new designs is usually limited. The question is: once we know what customers want, and once we know how many different designs we can afford, how should we select these designs?

What we do in this chapter. In this chapter, we describe a reasonable mathematical model within which we can find an optimal collection of design.

Towards a mathematical model. Let us denote the number of parameters needed to describe different designs by n. Then, each design can be characterized by an n-dimensional vector $x = (x_1, \ldots, x_n)$. Let us assume that the unit of different parameters are selected in such a way that a unit of each parameter represents the same difference for the user. Under this selection, it is reasonable to assume that the user's difference between two designs can be described by the Euclidean distance $d(x, y)$ between the corresponding vectors $x = (x_1, \ldots, x_n)$ and $y = (y_1, \ldots, y_n)$:

$$d(x, x') = \sqrt{\sum_{i=1}^{n} (x_i - y_i)^2}.$$

We have a large number of potential users. For each user, some design is ideal, and the farther we are from this ideal design, the less desirable this design is. For our purposes, we can simply identify each user with this ideal vector x.

There are usually very many users, each of these users can be characterized by a vector x. Ideally, we should record all these vectors, but in practice, it is reasonable to describe how many users are in different zones. In other words, a reasonable way to describe the users is to provide the distribution on the set of all possible designs that characterizes how popular different designs are. A natural way to describe a distribution of customers is to provide the population density $\rho_u(x)$ at different points x from the corresponding n-dimensional region. For this function, $\rho_u(x) \geq 0$ and the integral $\int \rho_u(x)\,dx$ is equal to the total number of potential customers.

Similarly, we can have a large number of engineered designs. So, instead of explicitly listing these designs, we can simply describe how many different designs are manufactured in different zones. Let us describe the corresponding design density by $\rho_m(x)$. Here, $\rho_m(x) \geq 0$ and

$$\int \rho_m(x)\, dx = D, \tag{5.3.1}$$

where D denotes the total number of designs.

If a manufacturer produces an ideal design, then the potential customer will buy it for sure. The larger the distance between the ideal and the actual designs, the less probable it is that the customer will purchase this design. Let $p(r)$ be the probability that a customer will purchase a design at distance r from the ideal one.

When the average density of the actual designs is $\rho_m(x)$, this means that in an area of linear size r and volume $V = r^n$, we have $\rho_m(x) \cdot r^n$ designs. So, we have one design in the area of size r for which $\rho_m(x) \cdot r^n = 1$. This equality leads to $r = \dfrac{1}{\sqrt[n]{\rho_m(x)}}$. So, around the point x, the probability that a customer buys a design is equal to $p(r) = p\left(\dfrac{1}{\sqrt[n]{\rho_m(x)}}\right)$. In the area of volume dx around the point x, there are $\rho_u(x)\, dx$ customers. Since the proportion $p(r)$ of them buys the design, the total number of customers in this area who purchased some design is equal to

$$\rho_u(x) \cdot p(r)\, dx = \rho_u(x) \cdot p\left(\frac{1}{\sqrt[n]{\rho_m(x)}}\right) dx.$$

Thus, the total number C of customers who bought our designs is equal to

$$C = \int \rho_u(x) \cdot p\left(\frac{1}{\sqrt[n]{\rho_m(x)}}\right) dx. \tag{5.3.2}$$

Our objective is to maximize the overall profit. Let s be our gain from selling a single unit. Then, by selling units to C customers, we gain the amount $C \cdot s$. Let d_0 be the cost of generating one design; then, by producing D designs, we spend the amount $D \cdot d_0$. If we subtract the expenses from the gain, we get the profit

$$M = C \cdot s - D \cdot d_0. \tag{5.3.3}$$

Resulting optimization problem. We are given the functions $\rho_u(x)$ and $p(r)$ and the values s and d. We need to select a function $\rho_m(x)$ for which the profit (5.3.3) is the largest possible, where the values C and D by using formulas (5.3.1) and (5.3.2). In other words, we need to optimize the following expression:

$$M = s \cdot \int \rho_u(x) \cdot p\left(\frac{1}{\sqrt[n]{\rho_m(x)}}\right) dx - d_0 \cdot \int \rho_m(x)\, dx. \tag{5.3.4}$$

Towards a solution. To solve the above optimization problem, we differentiate the objective function M by each unknown $\rho_m(x)$ and equate the resulting derivative to 0. Thus, we get

$$s \cdot \frac{1}{n} \cdot p'\left(\frac{1}{\sqrt[n]{\rho_m(x)}}\right) \cdot \frac{1}{\sqrt[n]{\rho_m(x)} \cdot \rho_m(x)} \cdot \rho_u(x) - d_0 = 0, \tag{5.3.5}$$

where $p'(r)$ is the derivative of $p(r)$. By moving d to the right-hand side, we get an equivalent formula

$$s \cdot \frac{1}{n} \cdot p'\left(\frac{1}{\sqrt[n]{\rho_m(x)}}\right) \cdot \frac{1}{\sqrt[n]{\rho_m(x)} \cdot \rho_m(x)} \cdot \rho_u(x) = d_0. \tag{5.3.6}$$

By dividing both sides by $s \cdot \frac{1}{n} \cdot \rho_u(x)$, we keep all the terms depending on the unknowns in the left-hand side and move all the known terms to the right-hand side:

$$p'\left(\frac{1}{\sqrt[n]{\rho_m(x)}}\right) \cdot \frac{1}{\sqrt[n]{\rho_m(x)} \cdot \rho_m(x)} = \frac{d_0 \cdot n}{s \cdot \rho_u(x)}. \tag{5.3.7}$$

Thus, for $z \stackrel{\text{def}}{=} \frac{1}{\sqrt[n]{\rho_m(x)}}$, we get an equation

$$p'(z) \cdot z^{n+1} = \frac{d_0 \cdot n}{s \cdot \rho_u(x)}. \tag{5.3.8}$$

Thus, if we denote by i the function which is inverse to $p'(z) \cdot z^{n+1}$, we get, for z, an explicit formula

$$z = i\left(\frac{d \cdot n}{s \cdot \rho_u(x)}\right) \tag{5.3.9}$$

Once we know $z = \frac{1}{\sqrt[n]{\rho_m(x)}}$, we can the reconstruct the desired density $\rho_m(x)$ as $\rho_m(x) = \frac{1}{z^n}$, i.e., as

$$\rho_m(x) = \frac{1}{\left(i\left(\frac{d_0 \cdot n}{s \cdot \rho_u(x)}\right)\right)^n}. \tag{5.3.10}$$

So, we arrive at the following solution to our original problem.

Solution. Let us form an auxiliary function $p'(z) \cdot z^{n+1}$, where $p'(z)$ denotes a derivative, and then form an inverse function $i(z)$ to this auxiliary function. In other words, we define $i(z)$ in such a way that $i(p'(z) \cdot z^{n+1}) = z$ for all z. Then, the optimal distribution $\rho_m(x)$ of designs can be described by the formula (5.3.10).

Comment. Similar arguments are used to select optimal sensor placements (Sect. 2.1) and in the optimal setting of cloud computing (Sect. 3.1).

5.4 Use of Knowledge in Maintenance: A Case Study

In this section, as a case study, we consider the important problem of software migration and modernization. Computers are getting faster and faster; the operating systems are getting more sophisticated. Often, these improvements necessitate that we migrate existing software to the new platform. In an ideal world, the migrated software should run perfectly well on a new platform; however, in reality, when we try that, thousands of errors appear, errors that need correcting. As a result, software migration is usually a very time-consuming process. A natural way to speed up this process is to take into account that errors naturally fall into different categories, and often, a common correction can be applied to all error from a given category. To efficiently use this idea, it is desirable to estimate the number of errors of different types. In this section, we show how imprecise expert knowledge about such errors can be used to produce very realistic estimates.

The results of this section first appeared in [21].

Computers are ubiquitous. Computers are ubiquitous. In many aspects of our daily life, we rely on computer systems: computer systems record and maintain the student grades, computer systems handle our salaries, computer systems record and maintain our medical records, computer systems take care of records about the city streets, computer systems regulate where the planes fly, etc.

Most of these systems have been successfully used for years and decades—and this is what every user wants, to have a computer system that, once implemented, can effectively run for a long time, without a need for time- and effort-consuming maintenance.

Need for software migration and modernization. No matter how successful a computer system, the time comes when there is a need to modernize it.

The main reason for such a need comes from the fact that computer systems operate in a certain environment; they are designed:

- for a certain computer hardware—e.g., with support for operations with data pieces ("words") of certain length,
- for a certain operating system,
- for a certain programming language,
- for a certain computer interface, etc.

Eventually, the computer hardware is replaced by a new one. While all the efforts are made to make the new hardware compatible with the old code, there are limits to that: every hardware or software feature that makes possible the use of old software inevitably slows down the new system and thus makes its use not as convenient for new users. Every computer upgrade requires a trade-off to balance the interest of old and new users. As a result, after some time, not all the features of the old system are supported. In such situations, it is necessary to adjust the software so that it will work on a new system. This process is called *software migration* or, alternatively, *software modernization*, and the software that needs such a migration is called *legacy software*; see, e.g., [22–26].

Software migration and modernization is difficult. At first glance, software migration and modernization sounds like a reasonably simple task. Indeed, the main intellectual challenge of software design is usually when we have to invent new algorithms, new techniques—because the previous techniques cannot solve the practical problem; in software migration and modernization, these techniques have already been invented.

However, anyone who has ever tried to upgrade a legacy system knows that it is not as easy as it may sound at first glance. It may have been easier if every single operation from the legacy code was clearly explained and justified. This abundance of explanatory comments is what we all strive for, but the actual software is far from this ideal. There is a strong competition between software companies; whoever releases the product first has a great advantage, and whoever is last risks bankruptcy. In such an environment, there is no time to properly document all the code. Moreover, comments are sometimes obscured or even deleted on purpose, so that competitors would not learn about the ideas that make this code efficient.

In search for efficiency, many "tricks" are added by programmers that take into account specific hardware, specific operating system—and when the hardware and/or operating system changes, these tricks can slow the system down instead of making it run more efficiently. For example:

- some old image processing systems utilized the existence of hardware supported operations with long inputs;
- in newer RISC systems, with limited number of hardware supported operations, processing of large inputs is no longer hardware supported, and thus, the resulting software becomes very slow.

People who need to migrate the legacy code do not know which parts of the code contain such tricks.

A typical legacy code is huge: the corresponding system has a million or more lines. As a result, when a user tries to run an old legacy code on a new system, the compiler will produce an astronomical number of error messages: 5,000 or even 10,000 is a typical number.

How migration and modernization are usually done. Usually, migration is done the hard way: a software developer looks into each and every error message, tries to understand what is the problem, and come up with a correction. This is a very slow and very expensive process: correcting each error can take hours, and the resulting salary expenses can run to millions of dollars.

There exist tools that try to automate this process by speeding up the correction of each individual error. These tools definitely help, they speed up the required time by a factor of two, three, or even ten, but still thousands of errors have to be handled individually.

Resulting problem: need to speed up migration and modernization. Since migration and modernization of legacy software is a ubiquitous problem, it is desirable to come up with ways to speed up this process.

In this section, we propose such an idea, and we show how expert knowledge can help in implementing this idea.

Our main idea. Our main idea is based on the fact that modern compilers do not simply indicate that there is an error, they usually provide a reasonably understandable description of the type of an error. For example:

- it may be that a program is trying to divide by zero,
- it may be that a program is trying to access an element of an array with a computed index which is beyond the original bounds, etc.

Some of these types of error appear in numerous places in the software. Our experience shows that in many such places, these errors are caused by the same problem in the code. So, instead of trying to "rack our brains" over each individual error, a better idea is to look at all the errors of the given type, and come up with a solution that would automatically eliminate the vast majority of these errors.

This idea is actually natural. Judging by the current practice, this idea sounds innovative in software migration. However, if one looks at it from the general viewpoint, one can see that this from the viewpoint of a general algorithmic development, this is a very general idea.

This is how most algorithms originated; let us give a few historical examples. In many case, people wanted to know what will happen if we merge two groups of objects together. If we have a group of 20 sheep and we merge it with a group of 12 sheep, how many sheep will we have? While sheep herders were solving this type of problems, cow herders were solving similar problems about cows: what if we merge a group of 20 cows and a group of 12 cows, how many cows will there be total? Later on they discovered that it is possible to find an algorithm that would add objects no matter whether they are sheep, cows, or plates.

Similarly in our case, instead of dealing with individual errors, we try to come up with a general approach that would enable us to handle all (or at last almost all) errors of a given type.

This idea only works if we have sufficiently many errors of a given type. Of course, this idea saves time only if we have enough errors of a given type. For example, if we only have two or three errors of some type, it is probably faster to eliminate these few errors one by one than to try to come up with a general solution that would include all these errors.

How many errors of different type there are? Need for an expert knowledge. To successfully implement this approach, we therefore need to be able to predict how many errors of different type we will encounter.

To the best of our knowledge, there are currently no well-justified software models that can predict these numbers. What we do have is many system developers who have an experience in migrating and modernizing software. It is therefore desirable to utilize their experience.

Since experts usually describe their experience not in precise terms, but by using imprecise ("fuzzy") words from natural language, it is reasonable to use the known precisiation techniques to transform this expert knowledge into precise terms—in particular, techniques developed in fuzzy logic; see, e.g., [27–29].

How to find possible types of errors: general idea. In order to apply the proposed approach, we need to first know the possible types of errors.

The type of each error is usually determined by the compiler. So, all we have to do is to run the compiler on the migrated software, and record all error messages that are generated. Extracting all such error messages is a straightforward text processing problems, which can be easily done by many existing text processing programs.

How to find possible types of errors: example. For example, under Unix, we can catch all error messages in a single file. This can be done, e.g., by the following straightforward Unix command:

```
gmake -f sourcefile.mak > compile.log
```

This command sends all the error messages—which by default would otherwise go to the screen—into the specially created file compile.log.

In C compilers, error messages usually start with keywords like `error` or `warning`. The first line of each error message contains this keyword followed by the type of the corresponding error. For example, this line can contain the following statement:

```
warning #2940-D: missing return statement at end of
    non-void function
```

As a result, the list of possible types of errors can be extracted from the `complile.log` file by using the following Unix command:

```
grep " warning ". complile.log | sort -u list >
    categories.txt
```

The `grep` command finds all the lines that contain the word `warning` in the file `compile.log`. The resulting list of lines is then sorted *uniquely* (i.e., duplicate lines are avoided), and the sorted list is placed into the new file `categories.txt`.

How many errors of each type do we have? Now that we know the most frequent types of errors we can use the standard text processing programs to count how many errors of each type we have.

Example. For example, we can upload the file with error messages into a spreadsheet program (e.g., into Excel) and use its counting features.

What we are trying to describe. Once we know how many errors of different types are there, a reasonable idea is to start with the errors of the most frequent type. Once we have learned how to deal with these errors, we should concentrate on errors of the second most frequent type, etc. After a few iterations, when all frequently repeated errors are eliminated, we reach a stage on which for each remaining type, there are so few errors of this type that it is easier to deal with these errors one by one.

From this viewpoint, what we would like to describe is how many errors there are of different types. We would like to know the number of errors n_1 of the most frequent type, the number of errors n_2 of the second most frequent type, etc. In general, we want to know the numbers $n_1, n_2, \ldots$, for which

$$n_1 \geq n_2 \geq \ldots \geq n_{k-1} \geq n_k \geq n_{k+1} \geq \ldots$$

Available expert knowledge. We know that for every level k, the number of errors n_{k+1} of the next level is somewhat smaller than the number of errors n_k of a given type.

Similarly, if we compare the number of errors n_k of a given type and the number of errors n_{k+2} of the level $k+2$, then we can also say that n_{k+2} is smaller than n_k—and that the difference between n_k and n_{k+2} is larger than the difference between n_k and n_{k+1}. We can make similar statements about the relation between n_k and n_{k+3}, etc.

How can we precisiate this idea—first approximation: idea. Let us start with the rule that n_{k+1} is somewhat smaller than n_k. By using the usual fuzzy control methodology:

- We first formulate what it means for two given values of n_k and n_{k+1} to be consistent with this rule. In fuzzy logic, this is obtained by describing, for every two possible values n_k and n_{k+1}, the degree to which the pair of n_{k+1} and n_k is consistent with this rule.
- For a given n_k, we then apply a defuzzification procedure and get a single estimate for n_{k+1} as a function of n_k:

$$n_{k+1} = f(n_k). \tag{5.4.1}$$

Comments.

- The exact form of the function $f(n)$ function depends on which membership function we used to describe the imprecise term "somewhat smaller" and on what defuzzification procedure we use.
- In principle, in addition to using the standard fuzzy methodology, we can use any other appropriate techniques to precisiate the dependence of n_{k+1} and n_k. For example, we can use interval-valued fuzzy techniques which often lead to a more accurate description of expert knowledge; see, e.g., [30–32].

Which function $f(n)$ should we select? To select an appropriate function $f(n)$, let us take into account that in many cases, a software package that needs migration consists of two (or more) parts. Because of this, we can estimate the number of errors of type $k + 1$ in two different ways:

- We can use the overall number $n_k = n_k^{(1)} + n_k^{(2)}$ of k-th type errors in both parts to predict the overall number n_{k+1} of the $(k + 1)$-th type errors. In this case, we get the following estimate:

$$n_{k+1} \approx f(n_k) = f(n_k^{(1)} + n_k^{(2)}).$$

- Alternatively, we can start with the numbers of errors $n_k^{(1)}$ and $n_k^{(2)}$ in each part, predict the values $n_{k+1}^{(1)}$ and $n_{k+1}^{(2)}$, and then add up these predictions. As a result, we get the following estimate:

$$n_{k+1} \approx f(n_k^{(1)}) + f(n_k^{(2)}).$$

It is reasonable to require that these two approaches lead to the same estimate, i.e., that we have

$$f(n_k^{(1)} + n_k^{(2)}) = f(n_k^{(1)}) + f(n_k^{(2)})$$

for all possible values of $n_k^{(1)}$ and $n_k^{(2)}$. In other words, for any two natural numbers a and b, we should have

$$f(a + b) = f(a) + f(b). \tag{5.4.2}$$

For $a = b = 0$, the formula (5.4.2) implies that $f(0) = 2f(0)$ and thus, $f(0) = 0$. For any other integer a, this formula implies that

$$f(a) = f(1) + \ldots + f(1) \ \ (a \text{ times}),$$

i.e., that

$$f(a) = c \cdot a, \tag{5.4.3}$$

where we denoted $c \stackrel{\text{def}}{=} f(1)$. Thus, the most appropriate function $f(n)$ is a linear function $f(n) = c \cdot n$.

Resulting dependence n_k. Substituting the linear function $f(n) = c \cdot n$ into the Eq. (5.4.1), we conclude that $n_{k+1} = c \cdot n_k$. Thus, $n_2 = c \cdot n_1$, $n_3 = c \cdot n_2 = c^2 \cdot n_1$, etc. One can easily see that for a general k, we get

$$n_k = n_1 \cdot c^{k-1},$$

i.e., that $n_k = \frac{n_1}{c} \cdot c^k$ and thus,

$$n_k = A \cdot \exp(-b \cdot k), \tag{5.4.4}$$

where $A \stackrel{\text{def}}{=} \frac{n_1}{c}$ and $b \stackrel{\text{def}}{=} -\ln(c)$.

How accurate is this estimate? To check how accurate is this estimate, we compared it with the actual number of errors of different types obtained when migrating a health-related C-based software package from a 32-bit to a 64-bit architecture.

In the following table, the number of errors of type $k =$ ab is stored:

- in the a-th column (which is marked ax),
- in its b-th row (which is marked marked xb).

For example, the number of errors of type $k = 23$ is stored:

- in the 2-nd column (which is marked 2x),
- in its 3-rd row (which is marked x3).

	0x	1x	2x	3x	4x	5x	6x	7x
x0	–	308	95	47	13	5	2	1
x1	7682	301	91	38	13	4	2	1
x2	4757	266	85	34	12	4	2	1
x3	3574	261	81	34	12	4	2	1
x4	2473	241	76	30	11	3	2	1
x5	2157	240	69	24	9	3	2	1
x6	956	236	58	21	8	3	2	1
x7	769	171	57	19	8	3	1	1
x8	565	156	50	17	8	2	1	1
x9	436	98	47	17	6	2	1	–

One can easily see that for $k \leq 9$, we indeed have $n_{k+1} \approx c \cdot n_k$, with $c \approx 0.65$–0.75. Thus, the above simple rule described the most frequent errors reasonably accurately.

However, starting with $k = 10$, the ratio n_{k+1}/n_k becomes much closer to 1. Thus, the one-rule estimate is no longer a good estimate.

Let us use two rules: an idea. We have just mentioned that if we only use one expert rule, we do not get a very good estimate for n_k. A natural idea is this to use two rules:

- in addition to the rule that n_{k+1} is somewhat smaller than n_k,
- let us also use the rule that n_{k+2} is more noticeably smaller than n_k.

In this case, once we know n_k and n_{k+1}, we can use the standard fuzzy methodology (or any other appropriate methodology) and get an estimate

$$n_{k+2} = f(n_k, n_{k+1}).$$

Which function $f(n_k, n_{k+1})$ should we use? Similarly to the one-rule case, once we take into account that the software package consists of two parts, we can estimate the number of errors of type $k+2$ in two different ways:

- We can use the overall numbers $n_k = n_k^{(1)} + n_k^{(2)}$ and $n_{k+1} = n_{k+1}^{(1)} + n_{k+1}^{(2)}$ of k-th and $(k+1)$-th type type errors in both parts to predict the overall number n_{k+2} of the $(k+2)$-th type errors. In this case, we get the following estimate:

$$n_{k+2} \approx f(n_k, n_{k+1}) = f(n_k^{(1)} + n_k^{(2)}, n_{k+1}^{(1)} + n_{k+1}^{(2)}).$$

- Alternatively, we can start with the numbers of errors $n_k^{(p)}$ and $n_{k+1}^{(p)}$ in each part p, predict the values $n_{k+2}^{(1)}$ and $n_{k+2}^{(2)}$, and then add up these predictions. As a result, we get the following estimate:

$$n_{k+2} \approx f(n_k^{(1)}, n_{k+1}^{(1)}) + f(n_k^{(2)}, n_{k+1}^{(2)}).$$

It is reasonable to require that these two approaches lead to the same estimate, i.e., that we have

$$f(n_k^{(1)} + n_k^{(2)}, n_{k+1}^{(1)} + n_{k+1}^{(2)}) = f(n_k^{(1)}, n_{k+1}^{(1)}) + f(n_k^{(2)}, n_{k+1}^{(2)})$$

for all possible values of $n_k^{(1)}$, $n_{k+1}^{(1)}$, $n_k^{(2)}$, and $n_{k+1}^{(2)}$. In other words, for any two four numbers $a \geq a'$ and $b \geq b'$, we should have

$$f(a+b, a'+b') = f(a, a') + f(b, b'). \quad (5.4.5)$$

Let us solve the corresponding functional equation. We want to find the value $f(x, y)$ for all $x \geq y$. By taking $a = a' = y$, $b = x - y$, and $b' = 0$, we conclude that

$$f(x, y) = f(y, y) + f(x - y, 0). \quad (5.4.6)$$

From the same formula (5.4.5), we can now conclude that

$$f(y, y) = f(1, 1) + \ldots + f(1, 1) \ \ (y \text{ times}),$$

i.e., that

$$f(y, y) = c_1 \cdot y \quad (5.4.7)$$

for a real number $c_1 \stackrel{\text{def}}{=} f(1, 1)$.

Similarly, from the property (5.4.5), we conclude that

$$f(z, 0) = f(1, 0) + \ldots + f(1, 0) \ \ (z \text{ times}),$$

i.e., that

$$f(z, 0) = c_2 \cdot z, \tag{5.4.8}$$

where $c_2 \stackrel{\text{def}}{=} f(1, 0)$.

Substituting the expression (5.4.7) and (5.4.8) into the formula (5.4.6), we conclude that

$$f(x, y) = c_1 \cdot y + c_2 \cdot (x - y) = c_2 \cdot x + (c_1 - c_2) \cdot y.$$

In other words, we conclude that $f(x, y)$ is a linear function of x and y. Thus, we have

$$n_{k+2} = a \cdot n_k + b \cdot n_{k+1} \tag{5.4.9}$$

for some constants $a \stackrel{\text{def}}{=} c_1 - c_2$ and $b \stackrel{\text{def}}{=} c_1$.

Resulting dependence n_k. Let us use the difference Eq. (5.4.9) to find the dependence of n_k on k. A general solution to a difference equation with constant coefficients is well known (see, e.g., [33]). In general, this solution is a linear combination of the expressions ρ^k, where ρ is a solution (real or complex) of the polynomial equation that is obtained when we plug in ρ_k into the corresponding difference equation. If the equation has a double or triple solution, then we can also consider the terms $k \cdot \rho^k$, $k^2 \cdot \rho^k$, etc.

In our case, substituting $n_k = \rho^k$ into the Eq. (5.4.9) and dividing both sides of the resulting equality by ρ^k, we conclude that

$$\rho^2 = a + b \cdot \rho. \tag{5.4.10}$$

This is a quadratic equation, and a quadratic equation either has two different real roots, or a single double real root, or it has complex conjugate roots.

For complex-conjugate roots ρ and ρ^*, the corresponding dependence has the following form:

$$n_k = A_1 \cdot \rho^k + A_2 \cdot (\rho^*)^k = A_1 \cdot (z \cdot k) + A_2 \cdot (z^* \cdot k) =$$

$$\text{const} \cdot \exp(p \cdot k) \cdot \cos(q \cdot k) + \text{const} \cdot \exp(p \cdot k) \cdot \sin(q \cdot k),$$

where $p + q \cdot \mathrm{i} = \ln(a + b \cdot \mathrm{i})$. This dependence contains trigonometric terms and is, thus, oscillating—and we want to a dependence for which always $n_k \geq n_{k+1}$.

So, in our case, the case of complex roots can be excluded, and we are left with situations in which we either have two different real roots, or one double real root. So, we have either

$$n_k = A_1 \cdot \rho_1^k + A_2 \cdot \rho_2^k, \tag{5.4.11}$$

or

$$n_k = A_1 \cdot \rho_1^k + A_2 \cdot k \cdot \rho_1^k, \tag{5.4.12}$$

for some values c_i and ρ_i. In other words, we have either

$$n_k = A_1 \cdot \exp(-b_1 \cdot k) + A_2 \cdot \exp(-b_2 \cdot k), \tag{5.4.13}$$

or

$$n_k = A_1 \cdot \exp(-b_1 \cdot k) + A_2 \cdot k \cdot \exp(-b_1 \cdot k), \tag{5.4.14}$$

where $b_i \stackrel{\text{def}}{=} -\ln(\rho_i)$.

With this new model, we get a much better fit with the data. Which of the models (5.4.13) and (5.4.14) is the best fit for the above data? One can see that the degenerate model (5.4.14) is close to exponential and thus, is not a good fit for the above experimental data.

So, we need to consider a general model (5.4.13). In this case, the values b_i are different. Thus, without losing generality, we can assume that $b_1 < b_2$. So, the desired estimate n_k is the sum of two terms:

- a slower-decreasing term $A_1 \cdot \exp(-b_1 \cdot k)$, and
- a faster-decreasing term $A_2 \cdot \exp(-b_2 \cdot k)$.

Under this assumption, what is the relation between the values A_1 and A_2?

If $A_1 > A_2$, then:

- for $k = 1$, the first term is larger, and
- since the second term decreases faster, the first term dominates for all k.

In this case, the expression (5.4.13) is close to an exponential function $A_1 \cdot \exp(-b_1 \cdot k)$, and we already know that an exponential function is not a good description of n_k.

Thus, to fit the empirical data, we must use models with $A_1 < A_2$. In this case:

- for small k, the second—feaster-decreasing—term $A_2 \cdot \exp(-b_2 \cdot k)$ dominates;
- however, since the second term decreases exponentially faster than the first one, for larger k, the first—slower-decreasing—term $A_1 \cdot \exp(-b_1 \cdot k)$ dominates.

Thus:

- for small k, we have $n_k \approx A_2 \cdot \exp(-b_2 \cdot k)$;
- for larger k, we have $n_k \approx A_1 \cdot \exp(-b_1 \cdot k)$.

In effect, we here have *two* exponential models:

- the first model works for small k, while
- the second model works for large k.

This double-exponential model indeed describes the above data reasonably accurately:

- for $k \leq 9$, as we have mentioned, the data is a good fit with an an exponential model for which $\rho = n_{k+1}/n_k \approx 0.65$–$0.75$;
- for $k \geq 10$, the data is a good fit with another exponential model, for which $\rho^{10} \approx 2$–3.

Practical consequences. Since for small k, the dependence n_k rapidly decreases with k, the values n_k corresponding to small k constitute the vast majority of all the errors. In the above example, 85 percent of errors are of the first 10 types. Thus, once we learn to repair errors of these type, the remaining number of un-corrected errors decreases by a factor of seven. This observation has indeed led to a significant speed-up of software migration and modernization.

Summary. In many practical situations, we need to migrate legacy software to a new hardware and system environment. Usually, if we simply run the existing software packages in the new environment, we encounter thousands of difficult-to-correct errors. As a result, software migration is very time-consuming. A reasonable way to speed up this process is to take into account that errors can be naturally classified into categories, and often all the errors of the same category can be corrected by a single correction.

Coming up with such a joint correction is also somewhat time-consuming; the corresponding additional time pays off only if we have sufficiently many errors of this category. So, to plan when to use this idea, it is desirable to be able to estimate the number of errors n_k of different categories k. In this section, we show that an appropriate use of expert knowledge leads to a double-exponential model (5.4.13) that is in good accordance with the observations.

References

1. O. Lerma, C. Tweedie, V. Kreinovich, Increased climate variability is more visible than global warming: a general system-theory explanation, in *Proceedings of the International Conference on Risk Analysis in Meteorological Disasters RAMD'2014*, Nanjing, China, 12–13 Oct, 2014
2. The Intergovernmental Panel on Climate Change (IPCC), *Fourth Assessment Report (AR4)*, 2007
3. The Intergovernmental Panel on Climate Change (IPCC), *Climate Change 2013: The Physical Science Basis*, IPCC Report, 2013
4. The Intergovernmental Panel on Climate Change (IPCC), Climate Change *Impacts* (Adaptation and Vulnerability, IPCC Report, 2014), p. 2014
5. The Intergovernmental Panel on Climate Change (IPCC), *Climate Change 2014: Mitigation of Climate Change*, IPCC Report, 2014
6. F.P. Kazakevich, A. M. Krapivin, Investigations of heat transfer and aerodynamical resistance in tube assemblies when the flow of gas is dustladen, *Izvestiya Vysshikh Uchebnykh Zavedenii Energetika*, no. 1 (1958), pp. 101–107 (in Russian)
7. W.T. Sproull, Viscosity of dusty gases. Nature **190**, 976–978 (1961)
8. P.G. Saffman, On the stability of laminar flow of a dusty gas. J. Fluid Mech. **13**, 120–128 (1962)
9. P.G. Saffman, *Vortex Dynamics* (Cambridge University Press, Cambridge, UK, 1995)
10. P.A. Shmitt, *Numerical and Analytical Analyses of a Tornado Model*, Master's Thesis, Department of Mathematical Sciences, Texas Tech University, Lubbock, Texas, 1999
11. V. I Shubov, Control of Vortex Flow and Tornado Models, in *Abstracts of the 1999 SIAM Annual Meeting*, Atlanta, Georgia, 12–15 May, 1999, Section MS65
12. V.I. Shubov, D.S. Gilliam, *Mathematical Analysis of Tornado Dynamics*, Technical Report, Department of Mathematical Sciences, Texas Tech University, Lubbock, Texas, 1999
13. K. Huang, *Introduction to Statistical Physics* (CRC Press, Boca Raton, Florida, 2001)

14. L.D. Landau, E.M. Lifschitz, *Statistical Physics* (Elsevier Butterworth-Heinemann, Oxford, UK, 2005)
15. V.N. Huynh, O. Lerma, V. Kreinovich, Kansei engineering: towards optimal set of designs, Int. J. Innov. Manag. Information & Production (IJIMIP) **3**(3), 49–53 (2012)
16. V.-N. Huynh, Y. Nakamori, and H. Yan, "A comparative study of target-based evaluation of traditional craft patterns using Kansei data", In: Y. Bi and M. A. Williams (eds.), *Proceedings of the 4th International Conference on Knowledge Science, Engineering and Management, KSEM'2010*, Belfast, Northern Ireland, UK, Sept 1–3, 2010, Springer Lecture Notes in Computer Science, Vol. 6291, 2010, pp. 160–173
17. M. Nagamachi, *Kansei/Affective Engineering* (CRC Press, Boca Raton, Florida, 2011)
18. M. Ryoke, Y. Nakamori, V.-N. Huynh, Personalized recommendation for traditional crafts using fuzzy correspondence analysis with Kansei data and OWA operator, in *Interval/Probabilistic Uncertainty and Non-Classical Logics*, ed. by V.-N. Huynh, Y. Nakamori, H. Ono, J. Lawry, V. Kreinovich, H.T. Nguyen (Springer Verlag, Berlin, Heidelberg, New York, 2008), pp. 311–325
19. H. Yan, V.-N. Huynh, T. Murai, Y. Nakamori, Kansei evaluation based on prioritized multi-attribute fuzzy target-oriented decision analysis. Info. Sci. **178**(21), 4080–4093 (2008)
20. H. Yan, V.-N. Huynh, Y. Nakamori, A group nonadditive multiattribute consumer-oriented Kansei evaluation model with an application to traditional crafts. Ann. Oper. Res. **195**(1), 325–354 (2012)
21. F. Zapata, O. Lerma, L. Valera, V. Kreinovich, How to speed up software migration and modernization: successful strategies developed by precisiating expert knowledge, in *Proceedings of the Annual Conference of the North American Fuzzy Information Processing Society NAFIPS'2015 and 5th World Conference on Soft Computing*, Redmond, Washington, 17–19 Aug, 2015
22. A. Menychtas, K. Konstanteli, J. Alonso, L. Orue-Echevarria, J. Gorronogoitia, G. Kousiouris, C. Santzaridou, H. Bruneliere, B. Pellens, P. Stuer, O. Strauss, T. Senkova, T. Varvarigou, Software modernization and cloudification using the ARTIST migration methodology and framework. Scalable Comput. Pract. Exp. **15**(2), 131–152 (2014)
23. A. Menychtas, C. Santzaridou, G. Kousiouris, T. Varvarigou, J. Gorronogoitia, O. Strauss, T. Senkova, L. Orue-Echevarria, J. Alonso, H. Bruneliere, B. Pellens, P. Stuer, ARTIST methodology and framework: a novel approach for the migration of legacy software on the cloud, in *Proceedings of the 15th IEEE International Symposium on Symbolic and Numeric Algorithms for Scientific Computing SYNASC'2013*, Timisoara, Romania, 23–26 Sept, 2013, pp. 424–431
24. I. Sahin, F. Zahedi, Policy analysis for warranty, maintenance, and upgrade of software systems. J. Softw. Maint.: Res. Pract. **13**, 469–493 (2001)
25. R.C. Seacord, D. Plakosh, G.A. Lewis, *Modernizing Legacy Systems: Software Technologies, Engineering Processes, and Business Practices* (Addison-Wesley, Boston, Massachusetts, 2003)
26. I. Warren, J. Ransom, Renaissance: a method to support software system evolution, in *Proceedings of the 26th Annual International Computer Software and Applications Conference COMPSAC'2002*, Oxford, UK, 26–29 Aug, 2002, pp. 415–420
27. G. Klir, B. Yuan, *Fuzzy Sets and Fuzzy Logic: Theory and Applications* (Prentice Hall, Upper Saddle River, New Jersey, 1995)
28. H.T. Nguyen, E.A. Walker, *First Course on Fuzzy Logic* (CRC Press, Boca Raton, Florida, 2006)
29. L.A. Zadeh, Fuzzy sets. Inf. Control. **8**, 338–353 (1965)
30. J.M. Mendel, *Uncertain Rule-Based Fuzzy Logic Systems: Introduction and New Directions* (Prentice-Hall, Upper Saddle River, New Jersey, 2001)
31. J.M. Mendel, D. Wu, *Perceptual Computing: Aiding People in Making Subjective Judgments* (IEEE Press and Wiley, Piscataway, New Jersey, 2010)
32. H.T. Nguyen, V. Kreinovich, Q. Zuo, Interval-valued degrees of belief: applications of interval computations to expert systems and intelligent control, *Int. J. Uncertain. Fuzziness Knowl. Based Syst. (IJUFKS)*, **5**(3), 317–358 (1997)
33. T.H. Cormen, C.E. Leiserson, R.L. Rivest, C. Stein, *Introduction to Algorithms* (MIT Press, Cambridge, Massachusetts, 2009)

Chapter 6
Conclusions

Abstract This section summarizes what we have done in this book.

For many decades, there has been a continuous progress in science and engineering applications. A large part of this progress comes from the new knowledge that researchers acquire, propagate, and use. This new knowledge has revolutionized many aspects of our life, from driving to communications to shopping.

However, there is one area of human activity which is the least impacted by the modern technological progress: the very processes of acquiring, processing, and propagating information. When we decide where to place sensors, which algorithm to use for processing the data—we rely mostly on our own intuition and on the opinion of the experts. As a result, knowledge-related methods that we select are often far from optimal. To make effective recommendations, it is necessary to build realistic models of the corresponding processes, and then use these models to find optimal ways of controlling these processes.

The need for such models is well understood. There are many numerical models of knowledge acquisition, processing, and propagations. Some of these models have been successfully used to enhance the corresponding processes. However, these applications are limited by the fact that most of these models are based on detailed numerical simulation of the corresponding processes, which make the resulting models very time-consuming to use. This is especially important in situations of *big data*, when the amount of data is so huge that the traditional numerical methods are not applicable. It is therefore necessary to develop *analytical* models for the corresponding knowledge-related processes, models that would allow easier optimization and application.

In this book, we have developed analytical models for all the knowledge-related processes, from knowledge acquisition to knowledge processing and knowledge propagation. We hope that our recommendations will be useful, especially for big data applications.

L.O. Lerma and V. Kreinovich, *Towards Analytical Techniques for Optimizing Knowledge Acquisition, Processing, Propagation, and Use in Cyberinfrastructure and Big Data*, Studies in Big Data 29, DOI 10.1007/978-3-319-61349-9_6

Index

L.O. Lerma and V. Kreinovich, *Towards Analytical Techniques for Optimizing Knowledge Acquisition, Processing, Propagation, and Use in Cyberinfrastructure and Big Data*, Studies in Big Data 29, DOI 10.1007/978-3-319-61349-9

Printed in the United States
By Bookmasters